Selected Titles in This Series

Volu

10

9

8

7

6

5

4

3

STUDENT MATHEMATICAL LIBRARY
Volume 10

The Mathematics of Soap Films: Explorations with Maple®

John Oprea

AMERICAN MATHEMATICAL SOCIETY

2000 *Mathematics Subject Classification.* Primary 49–01, 49–04, 49Q05, 53–01, 53–04, 53A10.

Maple is a registered trade mark of Waterloo Maple, Inc., Ontario, Canada

The figure on the cover was created by the author, using Maple.

Library of Congress Cataloging-in-Publication Data

Oprea, John.
  The mathematics of soap films : explorations with Maple® / John Oprea.
    p. cm. — (Student mathematical library, ISSN 1520-9121 ; v. 10)
  Includes bibliographical references and index.
  ISBN 0-8218-2118-0 (softcover : alk. paper)
  1. Minimal surfaces—Data processing.   2. Soap bubbles.   3. Surface tension.
4. Maple (Computer file)   I. Title.   II. Series.
  QA644.O67   2000
  516.3′62—dc21

                                                     00-041614

To Jan, Kathy and Bounder.

Contents

Contents

Preface

Scientific principles are often reflected in geometry. Whether it is the curve made by a hanging wire or the path that light takes around the sun, shapes are often the manifestations of Nature's design. This book is about the mathematics which describes the geometric properties of soap films. Using ideas from plane geometry, differential geometry, complex analysis and the calculus of variations, we can begin to understand why soap films take the shapes they do. But it isn't just soap which interests us as mathematicians. Rather, it is the mathematization of the study of soap film shapes which serves as a prime example of the place geometry has in mathematical modeling.

As we shall see, the effects of surface tension lead a soap film to minimize its surface area. This well-defined mathematical consequence allows us to study soap film shapes from a purely mathematical viewpoint. The mathematics involved ranges from the elementary to the very advanced, but in this book we will focus on a point somewhere in the middle. That is, readers are expected to know calculus and have some familiarity with differential equations, but the relevant notions from differential geometry and complex variables needed in the book are all discussed in Chapter 2. In fact, in order to get to the point quickly, we try to use only the essential ingredients of each of these subjects to begin to tell the mathematical story of soap films.

With this in mind, the book could serve as a text for a junior-senior level seminar course or an independent study course.

Basic references which vastly extend the exposition presented here are [Nit89], [DHKW92], [Oss86], [Boy59], [Mor88], [Tho92], [HT85] and [Ise92]. In particular, [Nit89] and [HT85] provide histories and overviews of the study of minimal surfaces (i.e. mathematical soap films) and the uses of minimal surfaces in science, engineering and architecture, while [Mor88] presents an introduction to geometric measure theory (the *modern* approach to minimal surfaces) which every budding geometer should read. For a straightforward discussion of surface tension and its effects, [Ise92] can't be beat.

In the past, minimal surfaces have been considered in differential geometry books as an add-on ([Gra93], [Opr97]) and in other books at perhaps too high a level for undergraduates. Also, the basics of surface tension have not been discussed in these books, but left to higher-level texts ([Fin86]) or books tending more toward the physics side of things ([Ise92]). So a major goal of this book is simply to make available a self-contained text where undergraduates can see a mixture of the different types of mathematics which pertain to minimal surfaces together with a bit of the science behind the subject. Most undergraduates only see the standard applications of mathematics to physical systems, where answers are typically only analytic. Here they can see another way that mathematics applies to science — the determination of optimal shape.

Since this book concentrates on shape, it would be a shame not to present the reader with ways of creating shapes. With this in mind, almost 40% of the book is devoted to exploring various notions using the software package Maple. So, in the last third of the book, the reader will 'see' fluids rising up inclined planes, create minimal surfaces from complex variable data and investigate the 'true' shape of a balloon. In fact, rather than reading the book in order, the reader is recommended to jump back and forth from the mathematical exposition to the relevant Maple work. It really pays nowadays for undergraduates to learn a package such as Maple or Mathematica, so the Maple work given is usually discussed at length. There is much to be learned in making computers actually do *interesting* things in mathematics.

With this in mind, I would like to thank John Reinmann for his expertise and interest in Maple applied to soap films, bubbles and, especially, variational problems. John's help has been invaluable. Also, my daughter Kathy has been my assistant in soap film demonstrations over the years and I would like to tell her here how much I appreciate that. In fact, the photographs of soap films in Chapter 1 were taken by Kathy Oprea and Katie Cline. These demonstrations and photographs are the outgrowth of a grant from Cleveland State University which enabled me to commission the sculptor Ron Dewey to create the wireframes needed for experimentation. Thanks to CSU and to Ron for his fine work. Finally, I'd like to thank my wife Jan for her support and for giving me my own small part of the house for my wireframes, a bucket of ever-ready soap solution and all the rest of my toys.

John Oprea
oprea@math.csuohio.edu
http://math3.math.csuohio.edu/~oprea

Chapter 1

Surface Tension

1.1. Introduction

All of us have had the experience of filling a glass with water which crests above the rim of the glass, but doesn't spill over. How can this happen? Think of a liquid as a collection of (polar) molecules which exert attractive forces of equal strength on each other. Deep inside the liquid a molecule feels equal forces from all directions, but near the surface of the liquid a molecule feels more of a force from inside the liquid than it does from the small number of molecules between it and the surface. Hence, those molecules near the surface are drawn into the liquid and the surface of the liquid displays a 'curvature'. This property of pulling the surface of a liquid taut is called *surface tension*. Thus we see liquids form into drops or bubbles because surface tension acts as a skin holding the liquid together. But a soap bubble doesn't shrink away to nothing, so at some point surface tension must be balanced by an internal pressure. In the case of the soap bubble, of course, the air pressure inside the bubble is larger than that outside, so an eventual equilibrium is attained. Nature often tells us when an equilibrium has been reached by making the geometry of the equilibrium state very special — and so it is with geometry constrained by surface tension. Excellent references for what follows are [**Ise92**], [**Lov94**] and [**Fin86**].

1.2. The Basics of Surface Tension

When two fluids are in contact with one another, the contact surface along which they meet displays behavior unlike either of the two fluids. As Thomas Young [**You05**] pointed out, the separating surface acts as a membrane between the fluids. In general, we will consider the situations where one of the two fluids is air or where the two fluids meet at a solid as well (i.e. capillarity).

The molecules in a fluid feel forces, generally electrical, from nearby molecules. Of course, some molecules are *polar*; that is, their bonding structure creates a geometry in which one side of the molecule contains a plus electrical charge while the other side contains a minus charge.

Example 1.2.1. Water molecules H_2O have a geometry which places the two hydrogen molecules (+ charge) on one side of the molecule and the oxygen molecule (− charge) on the other. Thus, water is highly polar.

Polar molecules attract each other in the usual fashion; plus charges attract minus charges and repel other plus charges. Therefore, a polar molecule in the interior of a fluid will surround itself with attracted molecules oriented to match electrical charges in this way. The molecules near the surface of a fluid have a different environment, however, and *this is what creates surface tension*. Molecules near the surface of a fluid feel forces from inside the fluid, but do not feel the same forces from outside the fluid. Therefore, on the interface between a liquid and air, say, the fluid molecules will (on average) be pulled into the fluid, decreasing both the surface area of the interface and the density of the fluid in the region of the interface. Water, because it is highly polar, has a strong tendency to minimize area.

A soap solution (see Figure 1) consists of water molecules and soap molecules. A soap molecule is formed from a metal salt of a long chain fatty acid molecule and becomes ionized in solution [**Ise92**]. Ordinary soap is a sodium salt of a fatty acid. A standard example is sodium stearate $C_{17}H_{35}COO^-Na^+$. In solution, the Na^+'s are free ions, while the geometry of $C_{17}H_{35}COO^-$ is such that COO^- forms a negatively charged head with a long hydrocarbon chain tail

$C_{17}H_{35}$. The polar COO^- is attracted to the hydrogen parts of water molecules and is said to be *hydrophilic*. The hydrocarbon chain $C_{17}H_{35}$ is not attracted to water molecules (and so is pushed away by the hydrophilic heads), so it is called *hydrophobic*. Near the surface of a soap solution, the hydrophobic tails stick out from the solution while the hydrophilic heads are pulled into the solution. The soap ions which accumulate near the surface in a monolayer are said to be *adsorbed* on to the surface.

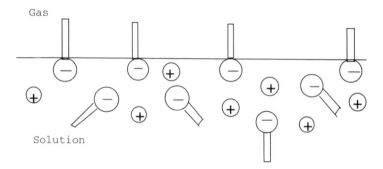

Figure 1: Hydrophilic and Hydrophobic Ions

A *soap film* (see Figure 2) consists of two such surfaces with a thin layer of fluid between them.

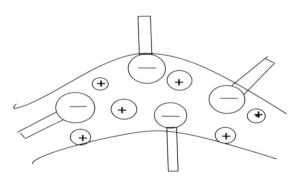

Figure 2: Ions in a Soap Film

The distances generally range from 2×10^5 Å to 50 Å (where Å is an *angstrom*, 1×10^{-10} meters). To put this into some perspective, this ranges from 50 times the wavelength of visible light to just a few atomic distances. The force which draws the interface inward is localized to within about one atomic thickness of the interface. The inward force produces a uniform tension across the surface. To see this, let's consider an old experiment. Take a horseshoe-shaped wire with a handle and with a rod across the horseshoe which is free to slide along the horseshoe. Dip the whole apparatus in a soap solution and remove it to reveal a soap film which pulls the slider towards the handle of the horseshoe.

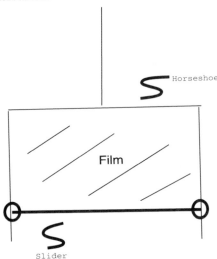

Figure 3: The Force of a Soap Film

Thus, the tension in the film produces a force acting on the slider. We see that the tension acts along the length of the sliding rod, so we give the tension units of force per unit length and define the *surface tension* to be

$$\sigma = \frac{F}{\ell}$$

where F is the force and ℓ is the length along which the force acts. There is another thing to notice too. We can hang a weight from the slider to exactly balance the force toward the handle, and no matter how far the slider is from the handle, the same weight works. This shows that soap films differ from a piece of rubber say, for the restoring force of stretched rubber becomes greater the more it is stretched. The reason for this discrepancy is that, unlike rubber, a stretched soap film replaces soap molecules on the surface with new ones from inside the liquid, thus keeping the density of soap molecules the same and, essentially, just creating more of the interface surface. Therefore, the tension and the force are kept constant as well. Here we should also mention that, to be precise, we should consider the length ℓ to be double the length of the slider since there are actually two sides to the film. We will think of this technicality as built into the definition of surface tension, rather than write 2ℓ.

The surface tension will be uniform, tangent to the surface and perpendicular to any line drawn in the surface interface. Furthermore, the inward force which creates surface tension also causes the interface surface to shrink in area as much as possible. Thus, *surface tension creates a tendency to minimize area.* (This is, in fact, equivalent to energy minimization here.) The units of surface tension are in dynes per centimeter (force per unit length). The following table [**Gar95**] lists some surface tension values for various fluids (where air is the other interface fluid). (Also see [**SZ47**].)

Liquid	Temp $C°$	Surface Tension (dynes/cm)
grain alcohol	20	26.8
glycerine	20	63.1
mercury	20	465.0
olive oil	20	32.0
soap solution	20	25.0
water	0	75.6
water	20	72.8
water	60	66.2
helium	-269	0.12

Soap solution has much lower surface tension than water because the non-polar hydrocarbon chains inhabit the interface between the solution and the air. If the soap film is perturbed (by shaking, say), then the surface area of the interface increases, decreasing the density of the hydrocarbon chains. Thus, the film acts more like water — the surface tension increases and either the film is forced to return to its equilibrium configuration or it breaks. This stabilizing effect is known as the *Marangoni effect*.

It is the lowering of surface tension that is responsible for the formation of soap films and bubbles. The surface tension of water is much too large for this to happen. Of course, this is also the reason that soap acts as a detergent. The lowering of surface tension allows dirt to be lifted from clothes or dishes. The following section provides some simple experiments illustrating some of the effects of surface tension.

1.3. Experiments with Soap Films

The following are some experiments which illustrate various soap film principles. For the reader who wishes to do these experiments, a good recipe for soap solution is provided by the San Francisco Exploratorium. The Exploratorium recommends using $\frac{2}{3}$ cup of Dawn and Joy dishwashing liquids together with 1 tablespoon of glycerine and 1 gallon of distilled water. A good source of other experiments is [Gar95]. There are many experiments in this wonderful little book, some of which we mention below. These experiments are, for the most part, easy to do with materials that are readily available. Some of the experiments we mention, however, are more conveniently done with special equipment.

1.3.1. The Paper Clip Experiment.

- Fill a plastic cup with water so that the water is even with the top of the cup. Add a few drops of water so that the water now rises *above* (surface tension!) the level of the cup. Set a paper clip on a fork and slowly lower the fork into the water. The paper clip will float on top of the water. Why?

- Now add one drop of soapy water to the cup. What happens? Why?

A Floating Paperclip

1.3.2. Drops on Wax Paper.

- Place a drop of water on a piece of wax paper. Look at the drop from the side and explain why you believe that water is not attracted to wax.

- Now add drops to the glob and observe the shape of the glob as it gets larger.

- Now, on the same sheet of wax paper, place drops of soapy water just as in the above two experiments. Explain what you see.

- Take a plastic basket used for strawberries and place it carefully so that it floats on water. Place a piece of paper towel in the basket. What happens? Now repeat the experiment, but place a piece of wax paper in the basket. Now what happens? Explain.

1.3.3. The Pepper Experiment.

- Sprinkle black pepper over the surface of water contained in a plastic cup. With an eyedropper, squeeze a drop of rubbing alcohol onto the surface of the water. What happens to the pepper? Explain.

- Observe for a few minutes. Now what happens to the pepper? Explain this. Add another drop of rubbing alcohol to the same place and observe what happens.

- Do the same experiment with soap solution in the eyedropper and observe similarities and differences with the alcohol experiment.

1.3.4. The Funnel and the Flame.

- Take a plastic funnel and dip the large end into soap solution to make a soap film. Blow into the small end of the funnel to make a bubble on the large end. Cover the small end so that the bubble remains.

- Light a candle, place the small end of the funnel near the flame and uncover the small end. What happens?

- Try this experiment with bubbles of different sizes. What do you observe about the size of bubbles and the strength of the effect?

1.3.5. Soap Motors.

- Cut out of aluminum foil a small boat shaped figure (i.e. tapered bow and flat stern).

- Fill a sink (or tub) with cold water and mix some soap solution near one side of the sink. Place the boat shape on the top of the water with the stern near the soapy side of the sink. What happens? Explain the propulsive power of soap!

1.3.6. Just Making Bubbles.

- Take a large bubble-blowing ring and a small bubble-blowing ring and blow some bubbles. What shape are the small bubbles? What shape are the large bubbles?

- Does the shape of the small bubbles change as they float through the air? How about the large bubbles? What could explain this?

1.3.7. The Horseshoe.

- Tie a piece of thread to the two ends of a wire shaped like a horseshoe (with handle). Leave a little slack in the thread so that it hangs down when you hold the horseshoe vertically.

- Dip the horseshoe in soap solution. What happens to the thread? § 1.4 has the answer. What shape do you think the thread makes? See Example 4.2.6 and Example 4.5.7.

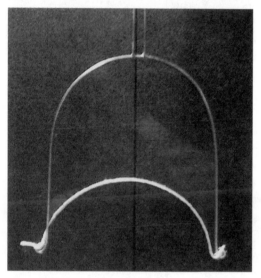

The Soap Film Horseshoe

1.3.8. Loop on a Soap Film.

- Take a standard plastic hoop with handle (commonly sold in stores for making bubbles) and dip it into soap solution to form a soap film across the hoop.

- Tie a piece of thread into a loop with a thread handle. Dip the loop into the soap solution and carefully lay the loop on the hoop's soap film. (You may need a second person for this. Don't forget to dip the thread before you place it on the film. Dry thread will pull so hard on the film that the film will break.)

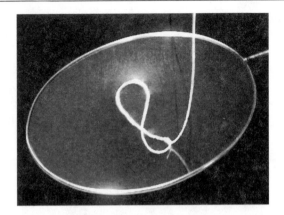

A Loop on a Soap Film

- Use a dry object such as a pencil to break the part of the soap film which is inside the loop of thread. Note that the surface tension of the film outside the loop pulls the loop into a *circle*. Why a circle? See Example 4.5.7.

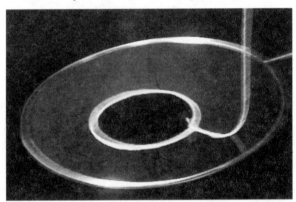

A Pierced Loop on a Soap Film

1.3.9. A Wound Wire.

- Take a reasonably flexible wire and wind it around a can to form a helix. Leave some room at the end for a handle.
- Dip the helix in a soap solution and remove it to find a soap film surface called a *helicoid*. The helicoid looks like a spiral

staircase, doesn't it? Why do you think this might be a good design?

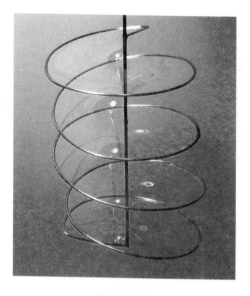

A Helicoid

1.3.10. Two Rings.

- Take two plastic rings (with handles), which are available from most toy stores, and dip them into a soap solution together. Take them out of the solution and have an assistant poke a (dry) finger through the center of the rings while keeping the rings together. There is now a soap film between the rings, but not inside each ring.

- Slowly pull the rings apart and watch the soap film between the rings. What happens? Does a cylinder form? The surface you are getting is called a *catenoid*. Can you pull the rings apart as far as you like and maintain the soap film?

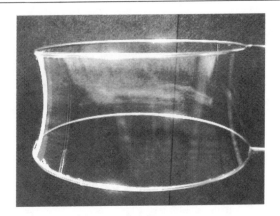

A Catenoid

- Have an assistant measure the separation of the rings when the film pops. Measure the diameter of the rings. Repeat the experiment and measurements with different size rings. Beware, this experiment may be affected by wind, humidity or even nerves! After experimenting, see § 5.6.

- Slowly pull the rings apart with the rings *not* concentric about a common axis. Now watch the soap film between the rings. The surface you are getting is called a *skew catenoid*. See the end of § 5.5 for a Maple picture.

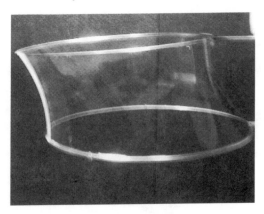

A Skew Catenoid

1.4. The Laplace-Young Equation

Let's analyze surface tension in the following way. Take a small piece of surface area expanded outward by an increased pressure as depicted in Figure 4. Since the piece is very tiny, we may suppose that the front and side curves are pieces of circles of radii R_1 and R_2 respectively. (This will be explained in Example 2.2.2.)

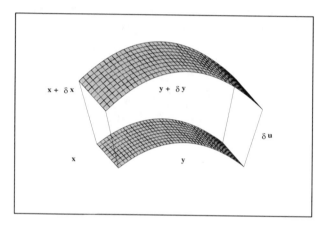

Figure 4: Expanding Surface Area

Let's compute the work done to expand the area (which we denote by S) given by a curvilinear rectangle with sides x and y (see Figure 4). Remember that work is force × distance, FD, and pressure p is force per unit area. We get

$$W = F \cdot D$$
$$= p \cdot S \cdot \delta u$$
$$= p \cdot x\, y \cdot \delta u.$$

Soap films have the potential to do work (see § 1.3.7), and we see that the amount of work done must be equal to the surface tension T (the units of which are force per unit length) multiplied by the change in surface area ΔS. This observation allows us to geometrize a physical problem. It is a physical principle that systems reach equilibrium when potential energy is minimized. Here, this energy is determined by the ability of the soap film to do work. In turn, the amount of work done is proportional to the change in surface area. Therefore, we have

Theorem 1.4.1 (First Principle of Soap Films). *A soap film takes a shape which minimizes surface area.*

Remark 1.4.2. These surface area minima may be local minima in the sense of calculus. Nevertheless, this property allows us to remove ourselves from the world of physical things and study soap films mathematically from the viewpoint of geometry.

Now let's return to the analysis of the expanding area. The change in area above is approximated by

$$\Delta S = (x + \delta x)(y + \delta y) - x\,y.$$

Now, the length x is part of a circle of radius R_1 and $x + \delta x$ is part of a circle of radius $R_1 + \delta u$ with *the same included angle*. Therefore, we have the equality

$$\frac{x + \delta x}{R_1 + \delta u} = \frac{x}{R_1}.$$

Treating y similarly gives

$$x + \delta x = x\left(1 + \frac{\delta u}{R_1}\right),$$

$$y + \delta y = y\left(1 + \frac{\delta u}{R_2}\right).$$

Plug these into ΔS to obtain

$$\Delta S = x\left(1 + \frac{\delta u}{R_1}\right) y\left(1 + \frac{\delta u}{R_2}\right) - xy$$

$$= x\,y\,\delta u\left(\frac{1}{R_1} + \frac{1}{R_2}\right) + x\,y\,\frac{(\delta u)^2}{R_1 R_2}.$$

If δu is small, we may neglect the last term and the resulting expression for ΔS may be plugged into the work formula $W = \sigma\Delta S$. We then end up with

$$p\,x\,y\,\delta u = \sigma\,x\,y\,\delta u\left(\frac{1}{R_1} + \frac{1}{R_2}\right)$$

$$p = \sigma\left(\frac{1}{R_1} + \frac{1}{R_2}\right).$$

This is the *Laplace-Young equation*, which was discovered around 1800 by T. Young and P. S. Laplace. It says that a pressure difference on either side of a bubble or film is given by the product of the surface tension σ of the bubble or film and a quantity which is clearly related to the shape of the bubble or film. In fact, notice that the only requirement (implicitly) placed on our infinitesimal piece of area was that the curves meet at right angles (so $S \approx xy$). These quantities $1/R_1$ and $1/R_2$ are the *normal curvatures* of the surface in those perpendicular directions (see Definition 2.2.1). In Chapter 2 we will see the quantity $2H = 1/R_1 + 1/R_2$ as twice the *mean curvature* of the surface. This derivation of the Laplace-Young equation is not particularly rigorous because of our assumptions about the shape of the piece of surface under consideration. More rigorous (but more difficult) derivations may be found in [**TF91**] (where it is attributed to Poisson) and [**Fin86**].

Exercise 1.4.3. Derive the Laplace-Young equation another way as follows. Consider the upper hemisphere of the R-sphere as a film in equilibrium. Compare forces in the z-direction. Show that the force due to surface tension is $2\pi R\sigma$ in the direction of the negative z-axis along the equator. Show that the force due to the pressure difference is $(p_{\text{under}} - p_{\text{above}})\pi R^2$ in the positive z-direction. Equilibrium says that the forces balance. (See [**SZ47**].)

1.5. Plateau's Rules for Soap Films and Consequences

J. Plateau studied soap films in the 1800's and empirically discovered three important rules for soap film formation (contained in a framework), the second of which was only rigorously proved in 1976 [**Tay76**] (see also [**AT76**]). These are:

I. *The 120° Rule.* Only three smooth surfaces of a soap film can meet along a line and the angle between any two of the three intersecting surfaces is 120°.

II.   *The* 109°28' *Rule.* Only four lines, each formed by the intersection of three surfaces, can meet at a point and the angle
between any pair of adjacent lines is arccos($-1/3$) $\approx$ 109°28'.

III. *The* 90° *Rule.* A soap film which is free to move along a
surface meets the surface at right angles.

We can't reproduce the proofs of these rules here because they
involve more involved techniques than we have available to us. The
proofs may be found in [**Tay76**], but there are informal discussions
in [**Lov94**] and [**HT85**]. We will, however, use these rules to derive
several interesting consequences. Also, the very fact that Plateau's
rules may be formulated takes us from the physical world to the world
of geometry. The argument below is a planar version of the 120° rule,
and arguments for the 90° rule are given in Example 3.9.13 and in
Example 4.3.7.

Plateau's Rules

Example 1.5.1 (Steiner's 3-Point Problem). Suppose three cities
are to be connected by a road system so that it is possible to get
from any city to any other and so that the total length of the road
system is a minimum. How can such a problem be solved? Here is
the *analogue* solution via soap films. We mentioned above that surface tension causes area to decrease to a minimum. If we accept this

principle for the moment, then we can achieve a length minimization if we keep the width of a film constant. To do this, enclose a soap film between two plexiglass sheets. Inside the sheets, place pins at locations corresponding to the locations of the cities. Dip the apparatus in soap solution and remove it. (If you dip vertically, you may have to tip the apparatus upside down to start the soap film flowing freely. Then tip horizontally for the final configuration.) The film that arises — a peace symbol without the circle — connects the cities (i.e. pins) and is of constant width. Hence, it minimizes length. Note that we are observing (and using) III (the 90° rule) here to ensure that the films are perpendicular to the plates. This analogue solution then provides the intuition behind a rigorous derivation.

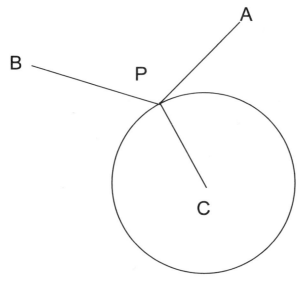

Figure 5: Steiner's Three-Point Problem

So, now let us give a geometrical proof (which can be found in various sources, e.g. [**Ise92**]). Let the three planar points be A, B and C (see Figure 5). First, note that, if these points lie on a line, then that line clearly solves the problem. So we assume A, B and C are the vertices of a triangle. Note that the minimum path cannot consist of curved lines since any such curved path can be replaced by

a straight line of shorter length (see Example 4.4.5). Secondly, take
any point Q outside the triangle $\triangle$ determined by vertices A, B and
C and draw straight lines to the vertices. Clearly, as long as Q is
outside the triangle, it is possible to move Q closer to $\triangle$ and reduce
the value of $AQ + BQ + CQ$. Hence, the minimum value point must
be inside the triangle if it exists.

Define a function $f \colon \triangle \to \mathbb{R}$ by $f(Q) = AQ + BQ + CQ$ where
$Q \in \triangle$ and AQ, BQ and CQ are the lengths of the line segments
joining the vertices of the triangle to Q. This function is continuous
and the triangle is a compact set (i.e. closed and bounded), so f
attains its maximum and its minimum. Of course, we are interested in
the minimum! Therefore, there exists a point P with $AP + BP + CP$
a minimum for all all points in $\triangle$. So, connect P by straight line
segments to A, B and C.

It's possible that P could be one of the vertices of the triangle.
Suppose A has the largest angle. Then the length $AB + AC$ is less
than the sum of the lengths of any other pair of sides of the triangle
since BC is the longest side. Therefore, we must have $P = A$. We will
see when this possibility actually arises later. For now, let's assume
that P is not one of the vertices.

Let $\mathcal{C} = \mathrm{circ}(C, CP)$ denote the circle of radius CP centered at
C. Note right away that A and B cannot be contained in this circle,
for if, for instance, A were inside, then

$$BP + CP + AP > BC + AC,$$

showing that the vertex C gives a minimum and contradicting the
minimality of $BP + CP + AP$. Now take ellipses with foci A and B
which intersect $\mathcal{C}$. Recall two facts about ellipses: if Q is a point on
the ellipse with foci A and B, then (1) $AQ + BQ$ is a constant $\mathcal{K}$ (for
that particular ellipse) and (2) the angle AQ makes with the tangent
to the ellipse at Q equals the angle BQ makes with the tangent (i.e.,
this is the reflection property). If one of the ellipses intersects the
circle in two points, then it can be shrunk to one with a smaller
value of $\mathcal{K}$. Only when the ellipse meets the circle in a single point
is this not possible. But there is an ellipse through P and P gives
the minimum of $BP + CP + AP$, so P also gives the minimum of

$BP + AP = \mathcal{K}$ since CP is fixed for all points on the circle $\mathcal{C}$. *So, we have found that P is given as the point of tangency of $\mathcal{C}$, the circle centered at C of radius CP, and the ellipse with foci A and B of smallest $\mathcal{K}$-value which intersects the circle.* The tangency condition says that the tangent line to the ellipse is the same as the tangent line to the circle $\mathcal{C}$, and so meets the segment CP (the radius of $\mathcal{C}$) at $90°$. Let the angle which AP (or BP) makes with the tangent line be θ and note that

$$\angle APC = 90° + \theta, \qquad \angle BPC = 90° + \theta.$$

Now we can carry out the procedure above for the other vertices. That is, we can take circles centered at A and B of respective radii AP and BP and shrinking ellipses with respective foci $\{B, C\}$ and $\{A, C\}$. We obtain angles ϕ and ρ with

$$\angle APC = 90° + \phi, \qquad \angle APB = 90° + \phi,$$
$$\angle APB = 90° + \rho, \qquad \angle CPB = 90° + \rho.$$

Let's compare these equations. Comparing the equations for $\angle APC$, we obtain $\theta = \phi$. Comparing those for $\angle CPB = \angle BPC$, we get $\theta = \rho$, so that all of these angles are equal. Now, using the first case above, we know that the angle which AP (or BP) makes with the tangent line is θ, so $\angle APB = 180° - 2\theta$. By the above, $\angle APB = 90° + \phi = 90° + \theta$, so $3\theta = 90°$, or $\theta = 30°$. Then, plugging this into all the equalities above, we get

$$\angle APB = 120°, \quad \angle APC = 120°, \quad \angle BPC = 120°.$$

This is the $120°$ rule. There are two points to clear up, however. One is this: the shrinking ellipse argument above only works if P is not on the line segment AB, for if it is, then the ellipses shrink to the line segment and $\theta = 0$. What prevents this? Suppose P is on the line segment and consider the segment CP. It must meet AB at $90°$ to make $AP + BP + CP$ a minimum. But then do the argument with the circle centered at A of radius AP and the ellipse with foci B and C tangent to the circle. The tangent line at P meets AB at $90°$ (since we have a circle), but so does CP! Thus, they are the same. This is impossible since CP is the segment to the focus C and tangent lines never go inside conics. This contradiction says that P is not between vertices of $\triangle$. Finally, it must be said when the case of the minimum value point P being at a vertex arises. *If one angle of the triangle $\triangle$ is $\geq 120°$, then P must be at a vertex.* The reason is this. Suppose P were inside and draw the lines to the vertices, where we assume the angle at C is $\geq 120°$. Then CP makes two angles and one, say PCA, is $\geq 60°$. But, by the work above for P inside, $APC = 120°$ and the triangle APC already has $\geq 180°$, a contradiction. Can this ever happen if all angles of $\triangle$ are less than $120°$?

Exercise 1.5.2. Answer the question of whether a vertex can give the minimum of $AP + BP + CP$ if all angles of $\triangle$ are $< 120°$. Hints: the construction of P above still works, but P may not give a minimum when compared to the vertices — it does give a minimum compared to everything else. Also, only focus on the vertex with largest angle, say B. Draw a line from A through P and extend it out so that a perpendicular may be drawn from B intersecting the extended AP at D. Show that the triangle BPD is a $30° - 60° - 90°$ right triangle and use what you know about the sides of such a triangle. Do the same thing extending CP and dropping a perpendicular to it from B. You should get equalities and inequalities like $\frac{1}{2}BP + AP < BA$ and $\frac{1}{2}BP + CP < BC$.

Exercise 1.5.3. Consider the three points $(-a, 0)$, $(a, 0)$ and $(0, b)$ and find the point (x, y) such that the sum of the distances from (x, y) to the three given points is a minimum. Show that $x = 0$ by symmetry and use calculus to show that $y = a/\sqrt{3}$. What does this say about the angles formed? Note that the value of b doesn't matter.

Exercise 1.5.4. What if four cities make a square and are to be joined by a minimizing road system? What should the final shape of the system be? Here is a soap film hint. Now find the exact shape.

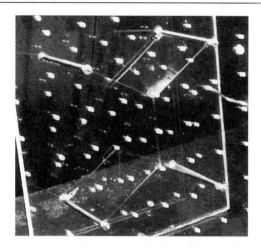

A Steiner 4-Point Problem

Example 1.5.5 (Two Fused Bubbles). Now let's consider two soap bubbles which have come together and fused (also see [**Ise92**], Theorem 1.7.4 and § 5.2). A planar section of this configuration looks like two circles intersecting with the arcs inside the circles erased and replaced by an arc of a third circle from one intersection point to the other which is the interface between the bubbles. (The fact that the interface is a circle follows from the constant pressure difference between the bubbles, the Laplace-Young equation and Alexandrov's theorem, Theorem 3.11.9.) Say that the original circles have centers at A and B with respective radii r_A and r_B (see Figure 6). (Without loss of generality, assume that $r_A > r_B$.) Let the top intersection point be O and let the center and radius of the interface circle be C and r_C. In the planar section, let X, Y and Z be points so that OZ is a tangent line to the bubble with center at A, OY is tangent to the circle centered at B and OX tangent to the circle centered at C.

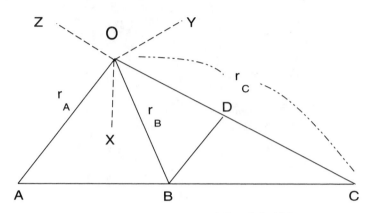

Figure 6: Cross-section of Fused Bubbles

By the $120°$ rule, OX, OY and OZ meet at $120°$. Since radii meet tangent lines of circles at $90°$, it is also true that OA and OC meet at $120°$. So, $AOZ = 90° = BOY$ and $AOC = 120°$. Further,

$$AOB = XOZ + XOY - AOZ - BOY$$
$$= 120° + 120° - 90° - 90°$$
$$= 60°.$$

Then, $BOC = 60°$ as well. Now construct a line through B which is parallel to OA and intersects OC at a point D. Because they have the same angles, the triangles AOC and BDC are similar. Thus,

$$\frac{BD}{AO} = \frac{DC}{OC}$$
$$DC = \frac{BD}{AO} OC$$
(†) $$= \frac{BD}{r_A} r_C.$$

Since BD is parallel to OA and OB is a transversal, then $OBD = AOB = 60°$. Now, $BOC = 60°$ also, so $ODB = 60°$ since the sum of a triangle's angles is $180°$. But then triangle OBD is an equilateral triangle, so $OD = BD = r_B$. Also, $DC = OC - OD = r_C - r_B$, so plugging into † gives

$$r_C - r_B = \frac{r_B}{r_A} r_C$$

$$r_A r_C = r_A r_B + r_B r_C$$

$$\frac{1}{r_B} = \frac{1}{r_C} + \frac{1}{r_A},$$

where the last line is obtained by dividing by $r_A r_B r_C$. This reciprocal relationship appears in many areas of applied mathematics. In particular, the total resistance (respectively, capacitance) of two resistors (respectively capacitors) in parallel is described by this relation. Thus, soap bubbles could be used as an analogue for such problems.

Exercise 1.5.6. Derive the reciprocal relation above by using the Laplace-Young equation and the obvious(?) relation between pressure differences between the bubbles and between the bubbles and the outside.

1.6. A Sampling of Capillary Action

All of us have seen liquids 'stick' to solid objects as the objects are removed from a solution. It's no surprise to us now that surface tension is the culprit here. In this brief section, let's look at this ability of liquids to 'rise up' from their ordinary equilibrium positions to climb the walls of planes and tubes.

Example 1.6.1 (A Liquid in Contact with an Inclined Plane). Suppose that a plane is set into a liquid at an angle β with respect to the horizontal (see Figure 7, where b is used instead of β, a instead of α and o instead of θ). The liquid rises up past the horizontal, attaching itself to the plane on both sides at a *contact angle* α intrinsic to the liquid and the solid which makes the plane. (The fact that the contact angle is a constant must be demonstrated. The best such demonstration, but the hardest, may be found in [**Fin86**]. Basic texts such as [**SZ47**] have more down to earth, but somewhat less convincing arguments.) The following discussion owes much to that in [**Ise92**].

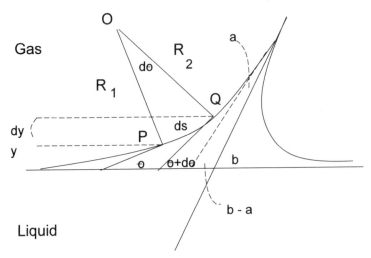

Figure 7: Liquid in Contact with an Inclined Plane

For instance, the contact angles of water-paraffin and mercury-glass combinations are 107° and 140° respectively. When $0° \leq \alpha \leq 90°$, then the liquid is said to *wet* the solid. Detergents decrease contact angle to allow wetting of solids. The pressures of the gas and liquid are determined by their densities ρ_g and ρ_ℓ, and the pressure difference p is given by ρgy, where $\rho = \rho_\ell - \rho_g$ is the density difference, g is the gravitational constant 980 cm/sec^2 and y is the height above the horizontal axis. The Laplace-Young equation is

$$p = \sigma \left(\frac{1}{R_1} + \frac{1}{R_2} \right)$$

and, along the horizontal direction of the plane, $R_2 = \infty$, since the liquid is straight in this direction. Therefore, for the inclined plane, the Laplace-Young equation becomes

$$\rho gy = \frac{\sigma}{R_1}.$$

Now take two close points P and (nearer the inclined plane) Q on the surface of the risen liquid and take the tangents to the interface at these points. These tangents intersect the horizontal in angles of θ and $\theta + \delta\theta$, where $\delta\theta$ is the angle between the normal segments OP

and OQ and O is the intersection of the normals of the tangent lines at P and Q. We think of the small portion of the interface between P and Q as lying on a circle of radius R_1, since this is what the principal curvature R_1 is meant to convey. Let δs denote the arc of the interface ($\approx$ circle) joining P to Q and note, by the arclength property of circles, that $R_1 \delta\theta = \delta s$. By making a small triangle with δs as hypotenuse, we can also approximate the vertical change in height δy by $\delta y = \delta s \sin\theta$. We then have

$$\rho g y = \frac{\sigma}{R_1} = \frac{\sigma \delta\theta}{\delta s} = \sigma \frac{\delta\theta}{\delta y} \sin\theta.$$

Taking limits, we obtain

$$\rho g y\, dy = \sigma \sin\theta\, d\theta$$

$$\rho g \int_0^y y\, dy = \sigma \int_0^\theta \sin\theta\, d\theta$$

$$\frac{1}{2}\rho g y^2 = \sigma(1 - \cos\theta)$$

$$= 2\sigma \sin^2\left(\frac{\theta}{2}\right),$$

with the result that the height y which the liquid rises up the plane is related to the angle θ at which the tangent intersects the horizontal by

$$y = 2\left(\frac{\sigma}{\rho g}\right)^{\frac{1}{2}} \sin\left(\frac{\theta}{2}\right).$$

Exercise 1.6.2. We know that a curve $y = y(x)$ has $dy/dx = \tan\theta$, where θ is the angle the tangent line makes with the horizontal axis. Use this to find a differential equation in x and y which may be solved to give y as a function of x. Hint: throw away constants to get $\theta = 2\sin(y)$ and then take $\tan(2\sin(y))$ and use trig identities to get the differential equation

$$\frac{dy}{dx} = \frac{2y\sqrt{1 - y^2}}{1 - 2y^2}$$

with solution (as given by Maple, see § 5.3)

$$\frac{1}{2} \operatorname{arctanh} \left(\frac{1}{\sqrt{1 - y(x)^2}} \right) - \sqrt{1 - y(x)^2} + x = C.$$

The highest point of the risen liquid has a tangent line which makes an angle of $\beta - \alpha$ with the horizontal, so the highest rise is

$$y_{\text{left}} = 2 \left(\frac{\sigma}{\rho g} \right)^{\frac{1}{2}} \sin \left(\frac{\beta - \alpha}{2} \right).$$

Exercise 1.6.3. Carry out the same analysis for the right side of the inclined (to the right at β) plane. Show that, essentially, the same argument works, but that the θ for the highest point is $\theta = 180° - (\alpha + \beta)$. Finally, show that

$$y_{\text{right}} = 2 \left(\frac{\sigma}{\rho g} \right)^{\frac{1}{2}} \cos \left(\frac{\alpha + \beta}{2} \right).$$

Exercise 1.6.4. Suppose a liquid rises up an inclined plane on the left by 4 mm with contact angle $\alpha = 30°$ and the plane is inclined with $\beta = 45°$. What is the surface tension σ? (Assume that the density difference ρ is 1 gm/cm^3.)

Example 1.6.5 (Rise of Liquid in a Capillary Tube). When a *very narrow* tube (called a *capillary tube*, from the Latin *capillus*, meaning 'a hair') is placed in a liquid, the level of liquid in the tube does not always match that of the ambient liquid (see Figure 8, where the contact angle is denoted by a). Water rises in a tube while mercury sinks. Why does this occur? It should be no surprise that surface tension is responsible for this phenomenon.

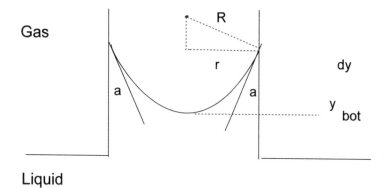

Figure 8: Rise of Liquid in a Capillary Tube

Let us see here (following [**Ise92**]) how what we already know can be used to predict the height of the capillary tube's liquid. The interface between the gas and the liquid is called the *meniscus*. First, again we use the Laplace-Young equation

$$p = \sigma \left(\frac{1}{R_1} + \frac{1}{R_2} \right)$$

to make a simplifying assumption. Recall that we may take $p = \rho g y$, where ρ is the density difference between the densities of the liquid and the gas, g is the gravitational constant and y is the height of the liquid. If the tube is very narrow, the y values throughout the meniscus change very little. Thus, $\rho g y$ is effectively constant. Hence, the mean curvature is also constant since the Laplace-Young equation says that $1/R_1 + 1/R_2$ is constant. Alexandrov's theorem (Theorem 3.11.9) implies that the meniscus is then part of a sphere. It is this simplifying approximation which allows the analysis of this case.

The meniscus meets the tube at a contact angle α in a circle of radius $R > r$, where r is the radius of the tube. (Assume for convenience that $\alpha \leq 90°$.) Therefore, the upward (i.e. vertical) force which draws the liquid upward to the surface is $2\pi r \sigma \cos \alpha$. A downward force is provided by the weight of the column of liquid under the interface. Suppose that the lowest point on the spherical meniscus is y_{bot} and that the highest point touching the tube is $y_{\text{bot}} + \delta y$.

The liquid below is cylindrical, so its volume is $\pi r^2 y_{\text{bot}}$ with weight $\rho g \pi r^2 y_{\text{bot}}$. A typical approximation here is to ignore the portion of the liquid above y_{bot} and obtain

$$\rho g \pi r^2 y_{\text{bot}} = 2\pi r \sigma \cos \alpha$$

$$y_{\text{bot}} = \frac{2\sigma \cos \alpha}{\rho g r}.$$

If we don't use the approximation, the volume of liquid from height y_{bot} to $y_{\text{bot}} + \delta y$ is obtained by subtracting the volume of the portion of the sphere bounded by the meniscus from the cylinder of radius r and height δy.

Exercise 1.6.6. To find the desired volume, follow these steps.

(1) Show that the radius of the sphere of which the meniscus is a part is $R = r/\cos \alpha$ and that $\delta y = r/\cos \alpha - r \tan \alpha$.

(2) Show that the volume of the portion of the sphere determined by the meniscus is

$$V_{\text{spherical part}} = \pi r^3 \left(\frac{2}{3} \sec^3 \alpha - \tan \alpha - \frac{2}{3} \tan^3 \alpha \right)$$

(3) and that the volume of liquid under the meniscus above level y_{bot} is

$$V_{\text{cyl} - \text{sph}} = \pi r^3 \left(\sec \alpha - \frac{2}{3} \sec^3 \alpha + \frac{2}{3} \tan^3 \alpha \right).$$

Thus, we have

$$2\pi r \sigma \cos \alpha = \rho g \pi r^2 \left(y_{\text{bot}} + r \left(\sec \alpha - \frac{2}{3} \sec^3 \alpha + \frac{2}{3} \tan^3 \alpha \right) \right).$$

We can now solve for the bottom point of the meniscus y_{bot} to get

$$y_{\text{bot}} = \left(\frac{2\sigma \cos \alpha}{\rho g r} \right) - r \left(\sec \alpha - \frac{2}{3} \sec^3 \alpha + \frac{2}{3} \tan^3 \alpha \right),$$

the first term of which is the approximation above. When the contact angle α is small, then we have the approximation

$$y_{\text{bot}} = \left(\frac{2\sigma}{\rho g} \right) \frac{1}{r} - \frac{r}{3}.$$

This can be useful in determining σ. Note however, that the assumption that r is very small cannot be done away with because it is this assumption which allows us to model the meniscus spherically. Also, a rigorous (and hard) derivation of this formula (as well as related items) by extremizing energy may be found in [**Fin86**]. For a Maple approach, see § 5.4.

1.7. Final Remarks

Example 1.7.1. In § 1.3.8, we saw that the pull of surface tension forces a loop to take the shape of a circle. By Theorem 1.4.1, surface tension tries to minimize the area of the soap film, possibly subject to a constraint such as having a fixed boundary. For the loop, the outside film minimizes, so the loop area maximizes when a circle forms. So, the loop experiment is like an analog computer which physically models the following

Theorem 1.7.2 (Meta-theorem). *The shape of the closed curve which maximizes its enclosed area subject to having a fixed perimeter is circular.*

Of course, we haven't proved this here, but we will later in Example 4.5.7! Nevertheless, what can we derive from the meta-theorem? Let's see. First, note that the circle of area A has

$$A = \pi R^2 = \pi (L/2\pi)^2 = \frac{L^2}{4\pi}$$

since the length L has $L = 2\pi R$. Thus, the meta-theorem says that it is always the case that any other curve with length L has area $A < L^2/4\pi$, or $2\sqrt{\pi A} < L$.

Now turn the problem around by asking for the closed curve with minimum length which surrounds a fixed area A. Of course, the circle has $L_{\text{circ}} = 2\pi R = 2\pi \sqrt{A/\pi} = 2\sqrt{\pi A}$. Suppose another curve encloses area A and has length $L \leq L_{\text{circ}}$. By what we said above, we have $2\sqrt{\pi A} < L$, but then we get

$$2\sqrt{\pi A} < L \leq L_{\text{circ}} = 2\sqrt{\pi A},$$

which is a contradiction. Therefore, we obtain

Theorem 1.7.3. *The shape of the closed curve which minimizes its length subject to enclosing a fixed area is circular.*

What is the analogue of this result in 3-dimensional space? Area is replaced by volume and length by surface area, so we might guess that the shape (now 2-dimensional) which minimizes surface area subject to enclosing a fixed volume is the 3-dimensional analogue of a circle — a sphere! But we know things which enclose volumes of air and which act to shrink surface area; soap bubbles. So, if the metatheorem is actually a theorem, then we would expect all soap bubbles to be spheres. What do you think? Blow some large and small bubbles to test your beliefs. Then see Example 3.11.4 and Example 4.5.10. Of course, there is no reason to stop at a single soap bubble. What about fused bubbles (see Example 1.5.5)? Surprisingly, it is only relatively recently that the following theorem has been proved.

Theorem 1.7.4 (Double Bubble Theorem [**HHS95**]). *A fused double bubble uniquely minimizes surface area among all surfaces in $\mathbb{R}^3$ enclosing two equal volumes.*

(In fact, in March 2000, it was announced by M. Hutchings, F. Morgan, M. Ritoré and A. Ros that the theorem holds without the hypothesis of *equal* volumes.) Here, the surface area includes the area of the disk separating the volumes. The very fact that this result took so long to prove underscores the difficulty of the subject. Nevertheless, at a non-expert level, there is much that still can be said, demonstrated and visualized in the study of the mathematics of soap films.

We have seen in this chapter a bit of the science and geometry underlying the formation of soap films. The one thing that strikes us over and over is this fundamental principle of minimization and the shapes it entails. In order to understand these shapes, we need some mathematical tools. In the next chapter, we will gather these tools by considering the fundamentals of differential geometry and complex variables in a brief and direct way. Then we shall be able to understand the geometry which can arise from the action of surface tension and the Universe's penchant for economy.

Chapter 2

A Quick Trip through Differential Geometry and Complex Variables

2.1. Parametrized Surfaces

The mathematics of soap films rests on the foundations of three subjects: differential geometry, complex analysis and the calculus of variations. In this chapter, we will review some of the basics of differential geometry and complex variables. In doing this, we will take our first steps toward mathematizing the physical problem of understanding soap films.

We see physical models of surfaces everywhere we go. The tubes inside the tires on our cars are tori. The bases of wicker chairs are (often) hyperboloids of one sheet. In order to study soap films mathematically, we need to describe them in mathematical terms; and this is our first order of business. Let's start with a familiar example. Consider a sphere of radius R with equation $x^2 + y^2 + z^2 = R^2$. Just because the sphere sits in $\mathbb{R}^3$, we shouldn't think of it as 3-dimensional. The surface of the Earth is a sphere and we know exactly where we are on it by knowing our latitude and longitude. A surface which can be described with two parameters must be 2-dimensional. Similarly, if we cut a paper towel cylinder lengthwise and unroll it to lie flat on a table, we see that the cylinder is really 2-dimensional. Therefore, when we want to understand a surface, we require two parameters (or coordinates) to describe it. Let's see how to do this.

Let D denote an open set in the plane $\mathbb{R}^2$. The open set D will typically be an open disk or an open rectangle. Let

$$\mathbf{x}\colon D \to \mathbb{R}^3, \qquad (u, v) \mapsto (x^1(u, v),\ x^2(u, v),\ x^3(u, v)),$$

denote a mapping of D into 3-space. This mapping is called a *parametrization* and the $x^i(u, v)$ are called component functions of the parametrization $\mathbf{x}$. We can do calculus one variable at a time on $\mathbf{x}$ by partial differentiation. Fix $v = v_0$ and let u vary. Then $\mathbf{x}(u, v_0)$ depends on one parameter and is, therefore, a curve. It is called a *u-parameter curve*. Similarly, if we fix $u = u_0$, then the curve $\mathbf{x}(u_0, v)$ is a *v-parameter curve*. Both curves pass through $\mathbf{x}(u_0, v_0)$ in $\mathbb{R}^3$, and tangent vectors for the u-parameter and v-parameter curves are given by differentiating the component functions of $\mathbf{x}$ with respect to u and v respectively. We write

$$\mathbf{x}_u = \left(\frac{\partial x^1}{\partial u},\ \frac{\partial x^2}{\partial u},\ \frac{\partial x^3}{\partial u} \right) \qquad \text{and} \qquad \mathbf{x}_v = \left(\frac{\partial x^1}{\partial v},\ \frac{\partial x^2}{\partial v},\ \frac{\partial x^3}{\partial v} \right).$$

We can evaluate these partial derivatives at (u_0, v_0) to obtain the tangent, or velocity, vectors of the parameter curves at that point, $\mathbf{x}_u(u_0, v_0)$ and $\mathbf{x}_v(u_0, v_0)$. Of course, to obtain true coordinates on a surface, we need two properties: first, $\mathbf{x}$ must be one-to-one (although we can relax this condition slightly to allow for certain self-intersections of a surface); secondly, $\mathbf{x}$ must never have $\mathbf{x}_u$ and $\mathbf{x}_v$ in the same direction because this destroys 2-dimensionality. The following lemma provides a simple test for the second property.

Lemma 2.1.1. *The tangent vectors $\mathbf{x}_u$ and $\mathbf{x}_v$ are linearly dependent if and only if $\mathbf{x}_u \times \mathbf{x}_v = 0$. (Recall that two vectors are linearly dependent if one is a scalar multiple of the other.)*

Proof. Recall that $\mathbf{x}_u \times \mathbf{x}_v$ is computed by a determinant

$$\det \begin{pmatrix} \mathbf{i} & \mathbf{j} & \mathbf{k} \\ x_u^1 & x_u^2 & x_u^3 \\ x_v^1 & x_v^2 & x_v^3 \end{pmatrix}.$$

Also, $\mathbf{x}_u$ and $\mathbf{x}_v$ are linearly dependent exactly when the last two rows of the determinant are multiples of each other. By the usual properties of determinants, this happens exactly when the determinant (and hence, the cross product) is zero.   □

Lemma 2.1.2. *Let M be a surface. If $\alpha\colon I \to \mathbf{x}(D) \subseteq M$ is a smooth curve in $\mathbb{R}^3$ which is contained in the image of a parametrization $\mathbf{x}$ on M, then for unique smooth functions $u(t)$, $v(t)\colon I \to \mathbb{R}$,*

$$\alpha(t) = \mathbf{x}(u(t), v(t)).$$

Proof. Take $\mathbf{x}^{-1}\alpha(t) = (u(t), v(t))$. Hence

$$\alpha(t) = \mathbf{x}(\mathbf{x}^{-1}\alpha(t)) = \mathbf{x}(u(t), v(t)).$$

In order to see that $u(t)$ and $v(t)$ are unique, suppose $\alpha = \mathbf{x}(\bar{u}(t), \bar{v}(t))$ for two other functions $\bar{u}$ and $\bar{v}$. Note that we can assume that $\bar{u}$ and $\bar{v}$ are defined on I as well by reparametrizing. Then

$$(u(t), v(t)) = \mathbf{x}^{-1}\alpha(t) = \mathbf{x}^{-1}\mathbf{x}(\bar{u}(t), \bar{v}(t)) = (\bar{u}(t), \bar{v}(t)).$$

□

So, to study a curve on a smooth surface, we can look at the functions of one variable $u(t)$ and $v(t)$. A vector $\mathbf{v}_p \in \mathbb{R}^3$ is *tangent to M at* $p \in M$ if $\mathbf{v}_p$ is the velocity vector of some curve on M. That is, there is some $\alpha\colon I \to M$ with $\alpha(0) = p$ and $\alpha'(0) = \mathbf{v}_p$. Usually we write $\mathbf{v}$ instead of $\mathbf{v}_p$. Then, the *tangent plane of M at p* is defined to be

$$T_p(M) = \{\mathbf{v} \mid \mathbf{v} \text{ is tangent to } M \text{ at } p\}.$$

The tangent vectors $\mathbf{x}_u$ and $\mathbf{x}_v$ are in $T_p(M)$ for $p = \mathbf{x}(u_0, v_0)$. The following result says that *every tangent vector is made up of a (unique) linear combination of $\mathbf{x}_u$ and $\mathbf{x}_v$*. Hence, $\{\mathbf{x}_u, \mathbf{x}_v\}$ form a *basis* for the vector space $T_p(M)$.

Lemma 2.1.3. *A vector $\mathbf{v}$ is in $T_p(M)$ if and only if $\mathbf{v} = \lambda_1 \mathbf{x}_u + \lambda_2 \mathbf{x}_v$, where $\mathbf{x}_u$, $\mathbf{x}_v$ are evaluated at (u_0, v_0).*

Proof. First suppose that α is a curve with $\alpha(0) = p$ and $\alpha'(0) = \mathbf{v}$. We saw above that $\alpha(t) = \mathbf{x}(u(t), v(t))$, so the chain rule gives $\alpha' =$

$\mathbf{x}_u(du/dt) + \mathbf{x}_v(dv/dt)$. Now, $\alpha(0) = p = \mathbf{x}(u(0), v(0))$, so $u(0) = u_0$ and $v(0) = v_0$ (since $\mathbf{x}$ is one-to-one) and

$$\mathbf{v} = \alpha'(0) = \mathbf{x}_u(u_0, v_0)\frac{du}{dt}(0) + \mathbf{x}_v(u_0, v_0)\frac{dv}{dt}(0).$$

Hence, $\lambda_1 = (du/dt)(0)$ and $\lambda_2 = (dv/dt)(0)$.

Now suppose $\mathbf{v} = \lambda_1\mathbf{x}_u + \lambda_2\mathbf{x}_v$ (where x_u and x_v are evaluated at (u_0, v_0)). We must find a curve on M with $\alpha(0) = p$ and $\alpha'(0) = \mathbf{v}$. Define this curve using the parametrization $\mathbf{x}$ by

$$\alpha(t) = \mathbf{x}(u_0 + t\lambda_1, \ v_0 + t\lambda_2).$$

Then $\alpha(0) = \mathbf{x}(u_0, v_0) = p$ and

$$\alpha'(t) = \mathbf{x}_u(u_0 + t\lambda_1, v_0 + t\lambda_2)\frac{d(u_0 + t\lambda_1)}{dt}$$
$$+ \ \mathbf{x}_v(u_0 + t\lambda_1, v_0 + t\lambda_2)\frac{d(v_0 + t\lambda_2)}{dt}$$
$$= \mathbf{x}_u(u_0 + t\lambda_1, v_0 + t\lambda_2)\lambda_1 + \mathbf{x}_v(u_0 + t\lambda_1, v_0 + t\lambda_2)\lambda_2.$$

Thus, $\alpha'(0) = \mathbf{x}_u\lambda_1 + \mathbf{x}_v\lambda_2$.   □

Example 2.1.4 (Parametrizations of Surfaces).

(1) *The Monge Parametrization.*

The graph of a real-valued function of two variables $z = f(x, y)$ is a surface in $\mathbb{R}^3$. To see this, define a parametrization by

$$\mathbf{x}(u, v) = (u, v, f(u, v)),$$

where u and v range over the domain of f. Then, $\mathbf{x}_u = (1, 0, \frac{\partial f}{\partial u})$ and $\mathbf{x}_v = (0, 1, \frac{\partial f}{\partial v})$. From now on, when convenient, we shall denote partial derivatives of functions by

$$\frac{\partial f}{\partial u} = f_u, \qquad \frac{\partial f}{\partial v} = f_v.$$

Also note that we never lose 2-dimensionality, since

$$\mathbf{x}_u \times \mathbf{x}_v = \det\begin{pmatrix} \mathbf{i} & \mathbf{j} & \mathbf{k} \\ 1 & 0 & f_u \\ 0 & 1 & f_v \end{pmatrix} = (-f_u, \ -f_v, \ 1) \neq 0.$$

(2) *Geographical Coordinates.*

Let M be a sphere of radius R (centered at $(0,0,0)$ for convenience). For a point a distance R away from the origin, draw a line segment from $(0,0,0)$ to the point. The parameter v (i.e. latitude), which varies as $-\pi/2 < v < \pi/2$, measures the angle (in radians) from this line down to the xy-plane. If the line is projected onto the xy-plane, then u (the longitude with $0 \leq u < 2\pi$) measures the angle of the projection from the positive x-axis. Hence, the xyz-coordinates of the point are

$$(R\cos u \cos v,\ R\sin u \cos v,\ R\sin v) = \mathbf{x}(u,v)$$

with

$$\mathbf{x}_u = (-R\sin u \cos v,\ R\cos u \cos v,\ 0),$$
$$\mathbf{x}_v = (-R\cos u \sin v, -R\sin u \sin v,\ R\cos v).$$

Further, we compute the cross product to be

$$\mathbf{x}_u \times \mathbf{x}_v = (R^2 \cos u \cos^2 v,\ R^2 \sin u \cos^2 v,\ R^2 \sin v \cos v),$$

where, in the third coordinate, we use $\sin^2 u + \cos^2 u = 1$. Note that $|\mathbf{x}_u \times \mathbf{x}_v| = R^2 \cos v$. From now on, we shall use the notation $S^2(R)$ for the sphere of radius R centered at the origin and simply S^2 for the sphere of radius 1 centered at the origin.

(3) *Surfaces of Revolution.*

Suppose C is a curve in the xy-plane and is parametrized by $\alpha(u) = (g(u), h(u),\ 0)$. Revolve this *profile curve* C about the x-axis. The coordinates of a typical point P may be found as follows. The x-coordinate is that of the curve itself, since we rotate about the x-axis. If v denotes the angle of rotation from the xy-plane, then the y-coordinate is shortened to $y\cos v = h(u)\cos v$ and the z-coordinate increases to $h(u)\sin v$. Hence, a parametrization may be defined by:

$$\mathbf{x}(u,v) = (g(u),\ h(u)\cos v,\ h(u)\sin v).$$

If we revolve about a different axis, then the coordinates permute. For example, a parametrized curve $(h(u), 0, g(u))$ in the xz- plane rotated about the z-axis gives a surface of revolution $(h(u)\cos v, h(u)\sin v, g(u))$. In general, $g(u)$ measures the distance along the axis of revolution and $h(u)$ measures the distance from the axis of revolution. The nicest situation is, of course, when $g(u) = u$, for then it is easy to see where you are on the surface in terms of the parameter u. The curves on the surface which are circles formed by revolving a single point about the axis are called *parallels* . The curves which look exactly like the original curve (but rotated out) are called *meridians*.

(4) *The Helicoid.*

Take a helix $\alpha(u) = (a\cos u,\ a\sin u,\ bu)$ and draw a line through $(0,0,bu)$ and $(a\cos u,\ a\sin u,\ bu)$. The surface swept out by this rising, rotating line is a helicoid. The line required is given by $(0,0,bu) + v(a\cos u, a\sin u, 0)$, so a parametrization for the helicoid is given by

$$\mathbf{x}(u,v) = (av\cos u, av\sin u, bu).$$

The helicoid is an example of a *ruled* surface.

Exercise 2.1.5. Find a parametrization for the *catenoid* obtained by revolving the catenary $y = \cosh(x)$ about the x-axis. The name *catenary* for the hyperbolic cosine will be explained in Example 4.2.5 and Example 4.5.5.

Exercise 2.1.6 (The Torus).

Revolve the circle of radius r centered at $(R,0,0)$ about the z-axis. Show that a parametrization is given by

$$\mathbf{x}(u,v) = ((R + r\cos u)\cos v,\ (R + r\cos u)\sin v,\ r\sin u).$$

How do u and v vary?

Exercise 2.1.7 (Enneper's Surface).

Define a surface by,

$$\mathbf{x}(u,v) = \left(u - \frac{u^3}{3} + uv^2,\ v - \frac{v^3}{3} + vu^2,\ u^2 - v^2\right).$$

Show that, for $u^2 + v^2 < 3$, Enneper's surface has no self-intersections. Hint: using polar coordinates $u = r\cos\theta$, $v = r\sin\theta$, show that the equality

$$x^2 + y^2 + \frac{4}{3}z^2 = \frac{1}{9}r^2(3 + r^2)^2$$

holds, where x, y and z are the coordinate functions of Enneper's surface, and then show that the equality implies that points in the (u, v)-plane on different circles about $(0, 0)$ cannot be mapped to the same point. For points with the same r and different θ_1, θ_2 (i.e. lying on the same circle), look at the third coordinate (in polar form) first and show that $\cos(2\theta_1) = \cos(2\theta_2)$. Then write the first and second coordinates as

$$r\cos\theta\left(1 + r^2 - \frac{4}{3}r^2\cos^2\theta\right), \quad r\sin\theta\left(1 + r^2 - \frac{4}{3}r^2\sin^2\theta\right)$$

to see that this would then imply that $\cos\theta_1 = \cos\theta_2$ and $\sin\theta_1 = \sin\theta_2$. Hence, $\theta_1 = \theta_2$. Finally, find two points on the circle $u^2 + v^2 = 3$ which *do* map to the same point of the surface. See § 5.8.

2.2. Normal Curvature

Because the cross product of two vectors is always perpendicular to each of the vectors themselves, we can always find a unit normal vector to the tangent plane of a surface M parametrized by $\mathbf{x}(u, v)$ by defining

$$U = \frac{\mathbf{x}_u \times \mathbf{x}_v}{|\mathbf{x}_u \times \mathbf{x}_v|}.$$

The *normal curvature in the tangent direction* $\mathbf{w}$ is defined as follows. Take the plane determined by the chosen unit direction vector $\mathbf{w}$ and the unit normal U, denoted $P = \text{Plane}(\mathbf{w}, U)$, and intersect this plane with the surface M. The intersection is a curve $\alpha(s)$ (which we can assume is unit speed). For unit speed curves, the curvature of the curve is detected by the acceleration of the curve (i.e. physically, think of Newton's second law), so the curvature in the *normal direction* should just be the projection of the acceleration onto the normal direction. In other words

Definition 2.2.1. For a unit tangent vector $\mathbf{w}$, the *normal curvature* in the $\mathbf{w}$-direction is defined to be

$$k(\mathbf{w}) = \alpha'' \cdot U,$$

where the derivatives are taken along the curve with respect to s.

Example 2.2.2. A circle of radius R has curvature $1/R$ because, for a parametrization $\alpha(s) = (R\cos(s/R), R\sin(s/R))$, we calculate $\alpha''(s) = (1/R)(-\cos(s/R), -\sin(s/R))$ and the Frenet formulas of curve geometry tell us that, since α is a unit speed curve, $\alpha'' = \kappa N$, where N is the unit normal of α. Comparing the formulas then shows that the curvature of α is $\kappa = 1/R$. For a surface M with normal U, if we take the plane P determined by U and a tangent vector $\mathbf{w}$ (at x_0) and let σ denote the curve formed by $P \cap M$, then it can be shown that $k(\mathbf{w}) = \kappa_{\sigma(0)}$ (where $\sigma(0) = x_0$). See [**Opr97**] for details. If we then take a small enough piece of σ about x_0, we can approximate σ near x_0 by a circle of radius $1/\kappa_{\sigma(0)}$. Hence, normal curvature may always be thought of as the reciprocal of the radius of some small circle which approximates the curvature of M in a particular direction. This justifies our approximations in the derivation of the Laplace-Young equation.

Lemma 2.2.3. *If α is a curve in M, then $\alpha'' \cdot U = -\alpha' \cdot U'$.*

Proof. We know that α' is a tangent vector and U is normal to the tangent plane, so $\alpha' \cdot U = 0$. Differentiate both sides of this equation to get

$$(\alpha' \cdot U)' = 0$$
$$\alpha'' \cdot U + \alpha' \cdot U' = 0$$
$$\alpha'' \cdot U = -\alpha' \cdot U'.$$

$\square$

Remark 2.2.4. Again, the derivatives are taken along the curve with respect to s. To take U', we again use the chain rule to get

$$U' = U_u \, u' + U_v \, v' \quad \text{with} \quad u' = \frac{du}{ds}, \quad v' = \frac{dv}{ds}.$$

We interpret $\alpha'' \cdot U$ as the component of acceleration due to the bending of M. Of course, we assume that α has unit speed so that the magnitude of α' does not affect our measurement. By Lemma 2.2.3 and Remark 2.2.4, we have

$$
\begin{aligned}
k(\mathbf{w}) &= -\alpha' \cdot U' \\
&= -(\mathbf{x}_u u' + \mathbf{x}_v v') \cdot (U_u u' + U_v v') \\
&= -\mathbf{x}_u \cdot U_u {u'}^2 - (\mathbf{x}_v \cdot U_u + \mathbf{x}_u \cdot U_v)u'v' - \mathbf{x}_v \cdot U_v {v'}^2 \\
&= \ell {u'}^2 + 2m u'v' + n {v'}^2,
\end{aligned}
$$

where the coefficients ℓ, m and n are given by

$$
(2.2.1) \quad \ell = -\mathbf{x}_u \cdot U_u, \quad 2m = -(\mathbf{x}_v \cdot U_u + \mathbf{x}_u \cdot U_v), \quad n = -\mathbf{x}_v \cdot U_v.
$$

These are the coefficients of the *second fundamental form*. The *first fundamental form* just describes how the surface distorts lengths from their usual measurements in $\mathbb{R}^3$. Namely, if α is a unit speed curve with tangent vector α', then $1 = |\alpha'|^2 = \alpha' \cdot \alpha'$ and we have

$$
\begin{aligned}
1 &= \alpha' \cdot \alpha' \\
&= (\mathbf{x}_u u' + \mathbf{x}_v v') \cdot (\mathbf{x}_u u' + \mathbf{x}_v v') \\
&= \mathbf{x}_u \cdot \mathbf{x}_u {u'}^2 + (\mathbf{x}_v \cdot \mathbf{x}_u + \mathbf{x}_u \cdot \mathbf{x}_v)u'v' + \mathbf{x}_v \cdot \mathbf{x}_v {v'}^2 \\
&= E {u'}^2 + 2F u'v' + G {v'}^2,
\end{aligned}
$$

where the coefficients E, F and G are called the (coefficients of) *the metric* and are given by

$$
E = \mathbf{x}_u \cdot \mathbf{x}_u, \quad F = \mathbf{x}_u \cdot \mathbf{x}_v, \quad G = \mathbf{x}_v \cdot \mathbf{x}_v.
$$

Now take any two perpendicular unit vectors $\mathbf{w}_1$ and $\mathbf{w}_2$ and find the normal curvature of each (denoted k_1 and k_2) using respective curves α_1 and α_2 with respective parameters $u_1(s)$, $v_1(s)$ and $u_2(s)$, $v_2(s)$. Then we have the equations (using $\mathbf{w}_1 \cdot \mathbf{w}_2 = \alpha_1' \cdot \alpha_2' = 0$ for the last)

$$
(2.2.2) \quad k_1 + k_2 = \ell({u_1'}^2 + {u_2'}^2) + 2m(u_1'v_1' + u_2'v_2') + n({v_1'}^2 + {v_2'}^2),
$$

$$
(2.2.3) \quad 1 = E{u_1'}^2 + 2F u_1'v_1' + G{v_1'}^2,
$$

$$
(2.2.4) \quad 1 = E{u_2'}^2 + 2F u_2'v_2' + G{v_2'}^2,
$$

$$
(2.2.5) \quad 0 = E u_1'u_2' + F(u_1'v_2' + u_2'v_1') + G v_1'v_2'.
$$

2.3. Mean Curvature

Using the results about normal curvature above, we can make precise
the type of curvature which plays the defining role in the study of
mathematical soap films.

Definition 2.3.1. Define the *mean curvature* function H by the relation

$$2H = k_1 + k_2,$$

where k_1 and k_2 are the normal curvatures associated to any two
perpendicular tangent vectors.

As it stands, the mean (or, *average*) curvature might not depend
only on the point of M, but on the particular directions chosen. We
will show that this is not the case by developing a formula for mean
curvature in terms of the coefficients E, F, G, ℓ, m and n alone. We
do this by using equations (2.2.2)-(2.2.5) above together with some
complex variable algebra (following [**Fin86**]).

We denote $\sqrt{-1}$ by i (so that $i^2 = -1$) and the field of *complex
numbers* by $\mathbb{C} = \{z = x + iy | x, y \in \mathbb{R}\}$. Complex numbers multiply
like polynomials. That is, if $z = x + iy$ and $w = a + ib$, then

$$\begin{aligned}
z \cdot w &= (x + iy) \cdot (a + ib) \\
&= xa + ixb + iya + iyib \\
&= xa - yb + i(xb + ya).
\end{aligned}$$

The *complex conjugate* of $z = x + iy$ is $\bar{z} = x - iy$. The *modulus* of
$z = x + iy$ is $|z| = \sqrt{x^2 + y^2} = \sqrt{z\bar{z}}$. So, with this bit of algebra in
mind, let's make the following definitions:

$$p = u_1' + iu_2' \quad \text{and} \quad q = v_1' + iv_2'.$$

Then we can compute

$$\bar{p} = u'_1 - iu'_2,$$
$$\bar{q} = v'_1 - iv'_2,$$
$$p\bar{q} = u'_1 v'_1 + u'_2 v'_2 + i(u'_2 v'_1 - u'_1 v'_2),$$
$$\bar{p}q = u'_1 v'_1 + u'_2 v'_2 - i(u'_2 v'_1 - u'_1 v'_2),$$
$$\frac{1}{2}(p\bar{q} + \bar{p}q) = u'_1 v'_1 + u'_2 v'_2,$$
$$pq = u'_1 v'_1 - u'_2 v'_2 + i(u'_2 v'_1 + u'_1 v'_2),$$
$$\bar{p}\bar{q} = u'_1 v'_1 - u'_2 v'_2 - i(u'_2 v'_1 + u'_1 v'_2),$$
$$p^2 = {u'_1}^2 - {u'_2}^2 + 2iu'_1 u'_2,$$
$$q^2 = {v'_1}^2 - {v'_2}^2 + 2iv'_1 v'_2.$$

From these calculations we get, from (2.2.2),

(2.3.1) $$2H = \ell p\bar{p} + m(p\bar{q} + \bar{p}q) + nq\bar{q},$$

and, by adding (2.2.3) and (2.2.4),

(2.3.2) $$2 = Ep\bar{p} + F(p\bar{q} + \bar{p}q) + Gq\bar{q}.$$

Furthermore, we have

$$
\begin{aligned}
Ep^2 + 2Fpq + Gq^2 &= E({u'_1}^2 - {u'_2}^2 + 2iu'_1 u'_2) + 2F(u'_1 v'_1 - u'_2 v'_2 \\
&\quad + i(u'_2 v'_1 + u'_1 v'_2)) + G({v'_1}^2 - {v'_2}^2 + 2iv'_1 v'_2) \\
&= 2i(Eu'_1 u'_2 + F(u'_1 v'_2 + u'_2 v'_1) + Gv'_1 v'_2) + E{u'_1}^2 \\
&\quad + 2Fu'_1 v'_1 + G{v'_1}^2 - (E{u'_2}^2 + 2Fu'_2 v'_2 + G{v'_2}^2) \\
&= 0 + 1 - 1 \\
&= 0
\end{aligned}
$$

by (2.2.3)-(2.2.5) above. Hence, we obtain the quadratic equation

$$E\left(\frac{p}{q}\right)^2 + 2F\left(\frac{p}{q}\right) + G = 0$$

having solutions

$$p = \left(-\frac{F}{E} + i\frac{\sqrt{EG - F^2}}{E}\right)q, \quad \bar{p} = \left(-\frac{F}{E} - i\frac{\sqrt{EG - F^2}}{E}\right)\bar{q}.$$

It is easy to compute that

$$p\bar{p} = \frac{G}{E}q\bar{q} \quad \text{and} \quad p\bar{q} + \bar{p}q = -\frac{2F}{E}q\bar{q}.$$

Now let's plug these quantities into (2.3.1) and (2.3.2) above. First, from (2.3.2) we obtain (after replacing $p\bar{p}$'s)

$$2 = Gq\bar{q} - \frac{2F^2}{E}q\bar{q} + Gq\bar{q} = \left(2G - \frac{2F^2}{E}\right)q\bar{q},$$

which gives

$$q\bar{q} = \frac{E}{EG - F^2}.$$

Now plug this into (2.3.1) for $q\bar{q}$ after replacing $p\bar{p}$ and $p\bar{q} + \bar{p}q$ to get

$$2H = \left(\ell\frac{G}{E} - m\frac{2F}{E} + n\right)q\bar{q}$$

$$= \left(\frac{G\ell - 2Fm + En}{E}\right)\left(\frac{E}{EG - F^2}\right)$$

$$= \frac{En + G\ell - 2Fm}{EG - F^2},$$

so that mean curvature is given by

$$H = \frac{En + G\ell - 2Fm}{2(EG - F^2)}.$$

Note that this formula *does not* depend on the original vectors $\mathbf{w}_1$ and $\mathbf{w}_2$, so any two perpendicular unit vectors may be chosen to compute mean curvature. In our derivation of the Laplace-Young equation, we took two perpendicular directions in the soap film surface, approximated the bounding curves by circles having radii R_1 and R_2 and expanded outward along the surface normal U. Because the curvature of a circle is simply the inverse of the radius (see Example 2.2.2), the normal curvatures for these bounding curves are $1/R_1$ and $1/R_2$. Therefore, we have

Corollary 2.3.2. *The Laplace-Young equation has the form*

$$p = 2\sigma H,$$

where H is the mean curvature of the film.

Corollary 2.3.2 is an indication that the following definition is the most important one in this book.

Definition 2.3.3. A surface M parametrized by $\mathbf{x}(u, v)$ is said to be a *minimal surface* if, at each point, $H = 0$.

2.4. Complex Variables

Although we saw above that complex variables can be used in even the most basic parts of differential geometry, a main point of this book is to show just how effective they are in the more advanced study of minimal surfaces.

Let's recall the basics of complex variable theory. As above, denote $\sqrt{-1}$ by i and the field of *complex numbers* by $\mathbb{C} = \{z = x + iy | x, y \in \mathbb{R}\}$. Let's recall some of the basic functions of complex analysis. For this, write $z = u + iv$ and define

$$e^z = e^u(\cos v + i \sin v)$$

and

$$\log(z) = \ln \sqrt{u^2 + v^2} + i \arctan\left(\frac{v}{u}\right).$$

We haven't been precise here about branches of the log, but this technicality will not concern us. Using the definition of e^z, we may define

$$\sin z = \frac{e^{iz} - e^{-iz}}{2i},$$

$$\cos z = \frac{e^{iz} + e^{-iz}}{2},$$

$$\sinh z = \frac{e^z - e^{-z}}{2},$$

$$\cosh z = \frac{e^z + e^{-z}}{2}.$$

One reason these definitions are chosen is that they extend the usual real functions of the same name. For example, if $z = u$, then the definition of e^z gives $\sin z$ as the real function $\sin u$. Similarly, for $z = u$, $\sinh z = \sinh u$. It is often useful to expand the complex functions into their real and imaginary parts. To accomplish this, we replace z by $u + iv$ and use the definition of e^z above.

Exercise 2.4.1. Derive the following formulas:

(1)　$\sin z = \sin u \cosh v + i \cos u \sinh v$.

(2)　$\cos z = \cos u \cosh v - i \sin u \sinh v$.

(3)　$\sinh z = \sinh u \cos v + i \cosh u \sin v$.

(4)　$\cosh z = \cosh u \cos v + i \sinh u \sin v$.

(5)　$\sin^2 z = (1 - \cos(2z))/2$.

Now let's turn away from the algebra and 'trigonometry' of complex numbers toward analysis in the complex plane. A function $f \colon \mathbb{C} \to \mathbb{C}$ is *continuous* at z_0 if $\lim_{z \to z_0} f(z) = f(z_0)$. If this is true for all z_0 in some open set D, then f is continuous in D. The function f is *complex differentiable* at $z_0 \in \mathbb{C}$ if

$$\lim_{z \to z_0} \frac{f(z) - f(z_0)}{z - z_0}$$

exists. In this case, the limit is denoted $f'(z_0)$. If the limit exists for all $z_0 \in D$, D open, then we say that f is *analytic* or *holomorphic* in D. Notice that, although this definition resembles the usual single variable calculus definition of derivative, it is much more subtle since z may approach z_0 from any direction along any kind of path. We can use this subtlety to our advantage, however. Because the range of f is $\mathbb{C}$, we can write $f(z) = f(x + iy) = \phi(x, y) + i\psi(x, y)$, where ϕ and ψ are real-valued functions of the two real variables x and y. The function ϕ is the *real part* of f while ψ is f's *imaginary part*. Now let's assume the limit above exists and compute it in the special case of $z = iy \to z_0 = iy_0$. We obtain

$$\lim_{y \to y_0} \frac{\phi(x_0, y) + i\psi(x_0, y) - [\phi(x_0, y_0) + i\psi(x_0, y_0)]}{i(y - y_0)}$$

$$= \lim_{y \to y_0} \frac{\phi(x_0, y) - \phi(x_0, y_0) + i\psi(x_0, y) - i\psi(x_0, y_0)}{i(y - y_0)}$$

$$= \lim_{y \to y_0} \frac{\phi(x_0, y) - \phi(x_0, y_0)}{i(y - y_0)} + \frac{i\psi(x_0, y) - i\psi(x_0, y_0)}{i(y - y_0)}$$

$$= \frac{1}{i}\frac{\partial\phi}{\partial y} + \frac{\partial\psi}{\partial y}$$

$$= \frac{\partial\psi}{\partial y} - i\frac{\partial\phi}{\partial y}.$$

Exercise 2.4.2. Show that, when $z = x \to z_0 = x_0$, then the limit is

$$\frac{\partial\phi}{\partial x} + i\frac{\partial\psi}{\partial x}.$$

Of course, if f is complex differentiable at z_0, then both of these limits are equal to $f'(z_0)$ and, hence, to each other. Therefore, we have

$$\frac{\partial\phi}{\partial x} = \frac{\partial\psi}{\partial y}, \qquad \frac{\partial\phi}{\partial y} = -\frac{\partial\psi}{\partial x}.$$

These are the *Cauchy-Riemann equations*. In fact, the analysis above may be enhanced to show that f is holomorphic on D if and only if $\frac{\partial\phi}{\partial x}, \frac{\partial\phi}{\partial y}, \frac{\partial\psi}{\partial x}, \frac{\partial\psi}{\partial y}$ exist and are continuous on D *and* the Cauchy-Riemann equations hold. We note here, without proof, that if f is holomorphic, so are all of its derivatives $f', f'', \ldots$.

Exercise 2.4.3. Show that $f(z) = z^2$ is holomorphic, and compute $f'(z)$ from the limit directly and from the Cauchy-Riemann equations.

Exercise 2.4.4. Compute the derivatives of the functions in Exercise 2.4.1.

Exercise 2.4.5. The *complex conjugate* of $z = x + iy$ is $\bar{z} = x - iy$. Show that $f(z) = \bar{z}$ is not holomorphic.

Exercise 2.4.6. The *modulus* of $z = x + iy$ is $|z| = \sqrt{x^2 + y^2} = \sqrt{z\bar{z}}$. If $f(z) = u + iv$ and $\bar{f}(z) = u - iv$, then we write $|f| = \sqrt{u^2 + v^2} = \sqrt{f\bar{f}}$. Show that a holomorphic function f with $|f|$ constant is itself constant.

Suppose f is holomorphic. The Cauchy-Riemann equations give

$$\frac{\partial^2 \phi}{\partial x^2} + \frac{\partial^2 \phi}{\partial y^2} = \frac{\partial}{\partial x}\frac{\partial \psi}{\partial y} - \frac{\partial}{\partial y}\frac{\partial \psi}{\partial x}$$
$$= \frac{\partial^2 \psi}{\partial x \partial y} - \frac{\partial^2 \psi}{\partial y \partial x}$$
$$= 0,$$

since mixed partials are equal. Thus, ϕ and (similarly) ψ satisfy the Laplace equation $\Delta \eta = 0$ (where $\Delta = \frac{\partial^2}{\partial x^2} + \frac{\partial^2}{\partial y^2}$).

Definition 2.4.7. A real-valued function $\Phi(x, y)$ is said to be *harmonic* if all of its second-order partial derivatives are continuous and on its domain it satisfies the Laplace equation,

$$\Delta \Phi \overset{\text{def}}{=} \frac{\partial^2 \Phi}{\partial x^2} + \frac{\partial^2 \Phi}{\partial y^2} = 0.$$

As we saw above, we then have the following.

Theorem 2.4.8. *If $f(z) = \phi(x, y) + i\psi(x, y)$ is holomorphic, then both $\phi(x, y)$ and $\psi(x, y)$ are harmonic.*

Example 2.4.9. Let $f(z) = z^2$, where $z = x + iy$. Then

$$z^2 = (x + iy)^2 = x^2 + 2ixy - y^2 = (x^2 - y^2) + i(2xy).$$

So, in this case, $\text{Re}(z^2) = \phi(x, y) = x^2 - y^2$ and $\text{Im}(z^2) = \psi(x, y) = 2xy$ are the particular harmonic functions produced from $f(z) = z^2$. The reader can check directly that ϕ and ψ are harmonic.

Exercise 2.4.10. Let $f(z) = z^3$, where $z = x + iy$. Calculate the real part of f. The harmonic function obtained produces the surface known as the monkey saddle.

To show just how special harmonic functions are, we state one of their salient properties. Let $U \subset \mathbb{R}^2$ be a bounded open set with closure $\overline{U}$ having boundary $\partial U = \overline{U} - U$.

Theorem 2.4.11. *Let $\phi \colon \overline{U} \to \mathbb{R}$ be a harmonic function which is continuous on $\overline{U}$ and differentiable on U. Then ϕ takes its maximum and minimum values on the boundary ∂U.*

Furthermore, if ϕ is a twice differentiable harmonic function of x and y, then on some open set there is another harmonic function ψ such that $f = \phi + i\psi$ is holomorphic. Harmonic functions ϕ and ψ which give such an f are said to be *harmonic conjugates*. As usual, for convenience, we shall often use the notation ϕ_x for the partial derivative with respect to x, etc.

Exercise 2.4.12. Let $\phi = x^2 - y^2$, and find its harmonic conjugate ψ as follows. From Cauchy-Riemann, $\phi_x = \psi_y$, so integrate ϕ_x with respect to y to get ψ up to a function of x alone. Now use $\phi_y = -\psi_x$ to determine this function. Are you surprised at your result?

Integration of complex functions is done by using line integrals from vector calculus. Suppose $f = \phi + i\psi$ is continuous and $\gamma(t) \colon [a, b] \to \mathbb{C}$ is a curve. Then we define the integral of f along γ to be

$$\int_\gamma f = \int_a^b f(\gamma(t))\gamma'(t)\, dt.$$

Exercise 2.4.13. Show that $\int_\gamma f = \int_\gamma \phi(x,y)\, dx - \psi(x,y)\, dy + i\int_\gamma \phi(x,y)\, dy + \psi(x,y)\, dx$, where the integrals on the right-hand side are real variable line integrals.

Exercise 2.4.14. Suppose f is holomorphic with a continuous derivative on and inside a closed curve γ. Use Green's theorem on each integral in the exercise above to show that $\int_\gamma f = 0$. This is a weak version of Cauchy's theorem. Show that this implies that integrals of holomorphic functions only depend on the endpoints and not on the paths chosen over which to integrate.

The most important thing for *us* to remember is that there is a Fundamental Theorem of Calculus for complex integrals. Namely,

Theorem 2.4.15. *If f is holomorphic, then*

$$\int_\gamma f' = f(b) - f(a).$$

Proof. Let $f(z) = \rho(u, v) + i\,\tau(u, v)$, so that $f'(z) = \rho_u + i\,\tau_u$. Also, let $\gamma(t) = u(t) + i\,v(t)$ be a curve in the complex plane with $a \leq t \leq b$. Then, using the Cauchy-Riemann equations $\rho_u = \tau_v$, $\rho_v = -\tau_u$ and the usual Fundamental Theorem of Calculus, we have

$$\int_\gamma f' = \int_a^b f'(\gamma(t))\gamma'(t)\,dt$$

$$= \int_a^b (\rho_u + i\,\tau_u)(u' + i\,v')\,dt$$

$$= \int_a^b \rho_u u' - \tau_u v' + i\,(\tau_u u' + \rho_u v')\,dt$$

$$= \int_a^b \rho_u u' + \rho_v v' + i\,(\tau_u u' + \tau_v v')\,dt$$

$$= \int_a^b \frac{d\rho}{dt} + i\frac{d\tau}{dt}\,dt$$

$$= \rho(u(b), v(b)) + i\tau(u(b), v(b)) - \rho(u(a), v(a)) - i\tau(u(a), v(a))$$

$$= f(b) - f(a).$$

$\square$

The last line, of course, really means that f is evaluated at the end-points of γ, but it is simpler to denote that as shown. Therefore, since there is a fundamental theorem of calculus, many of the formulas from ordinary calculus carry over into complex analysis. This will allow the calculation of complex integrals in the Weierstrass-Enneper representation later.

With a view to the future, when we shall consider a parametrization $\mathbf{x}(u, v)$ with complex coordinates, we write $z = u + iv$, $\bar{z} = u - iv$

and introduce the following notation for complex partial differentiation:

$$\frac{\partial}{\partial z} = \frac{1}{2}\left(\frac{\partial}{\partial u} - i\frac{\partial}{\partial v}\right), \qquad \frac{\partial}{\partial \bar{z}} = \frac{1}{2}\left(\frac{\partial}{\partial u} + i\frac{\partial}{\partial v}\right).$$

One advantage of this notation is that it provides an easy test for f to be holomorphic.

Exercise 2.4.16. Show that f is holomorphic if and only if $\frac{\partial f}{\partial \bar{z}} = 0$.

Exercise 2.4.17. Show that

$$\Delta f \stackrel{\text{def}}{=} f_{uu} + f_{vv} = 4\left(\frac{\partial}{\partial z}\left(\frac{\partial f}{\partial \bar{z}}\right)\right).$$

We will also need the following notion when we discuss the Weierstrass-Enneper representations. A complex function g is *meromorphic* if all its singularities (e.g. where the function is not defined) are poles (see [**MH87**] or [**SS93**]). That is, around each singularity z_0 there is a Laurent expansion (generalizing the Taylor expansion) of the form

$$g(z) = \frac{b_n}{(z - z_0)^n} + \cdots + \frac{b_1}{(z - z_0)} + \sum_{j=0}^{\infty} a_j(z - z_0)^j$$

for some finite n with coefficients determined by g. For us, the most important examples of meromorphic functions are rational functions

$$g(z) = \frac{\mathcal{P}(z)}{\mathcal{Q}(z)},$$

for polynomials $\mathcal{P}$, $\mathcal{Q}$. Then, the roots of $\mathcal{Q}$ are the singularities and, if not cancelled by roots of $\mathcal{P}$, are poles.

A much deeper theorem (see [**MH87**]) than any we have recalled so far is the following result which will be the key to solving Björling's problem in § 3.9. This is a case where complex variable theory differs radically from real variable theory, thus providing us with extra tools to create and study minimal surfaces.

Theorem 2.4.18 (The Identity Theorem). *If f and g are two holomorphic functions on a connected, open region $D \subseteq \mathbb{C}$ and $f(z_i) = g(z_i)$ for some convergent sequence $z_1, \ldots, z_n, \ldots \to \bar{z}$ in D, then $f = g$ on all of D.*

Exercise 2.4.19. Show that the function $f\colon \mathbb{R} \to \mathbb{R}$ defined by

$$f(x) = \begin{cases} x^2 \sin(2\pi/x), & x \neq 0, \\ 0, & x = 0. \end{cases}$$

is (real) differentiable at every point in $\mathbb{R}$ and that $f(1/n) = 0$ for all integers n. So differentiable real functions don't satisfy an identity theorem. Also, show that f' is not continuous at 0, so real functions differ from complex analytic ones which have the property that derivatives of all orders exist and are continuous.

2.5. Gauss Curvature

Although we shall primarily be concerned with mean curvature in this text, it is important to realize that yet another 'curvature' is equally important. This new curvature is called the *Gauss curvature* and is denoted by K. Here we shall simply develop one formula for the Gauss curvature.

Let's begin by using a simple trick which appears over and over in geometry. The unit normal U has the property $U \cdot U = 1$, so let's differentiate both sides of this equation using the *product rule* on the left. We get

$$(U \cdot U)_u = U_u \cdot U + U \cdot U_u = 2U_u \cdot U = 0$$

since the derivative of a constant is zero. Similarly, we may take the v-partial and we have $U_v \cdot U = 0$. Of course, all vectors which are perpendicular to the normal are tangent vectors. Hence, by Lemma 2.1.3, we have

$$(2.5.1) \qquad U_u = a\mathbf{x}_u + b\mathbf{x}_v \quad \text{and} \quad U_v = c\mathbf{x}_u + d\mathbf{x}_v.$$

In order to determine the coefficients a, b, c and d in (2.5.1), we can take dot products with the basis vectors $\mathbf{x}_u$ and $\mathbf{x}_v$. For instance, by (2.2.1),

$$-\ell = \mathbf{x}_u \cdot U_u = a\mathbf{x}_u \cdot \mathbf{x}_u + b\mathbf{x}_v \cdot \mathbf{x}_u = aE + bF.$$

Similarly, we obtain the following list:

$$\ell = -aE - bF,$$
$$m = -aF - bG,$$
$$= -cE - dF,$$
$$n = -cF - dG.$$

(For the middle two equations, we use $\mathbf{x}_u \cdot U = 0$ and the product rule to obtain $\mathbf{x}_{uv} \cdot U + \mathbf{x}_u \cdot U_v = 0$. Similarly, $\mathbf{x}_{uv} \cdot U + \mathbf{x}_v \cdot U_u = 0$. Then

$$2m = -(\mathbf{x}_v \cdot U_u + \mathbf{x}_u \cdot U_v)$$
$$= -2\mathbf{x}_v \cdot U_u$$
$$= -2\mathbf{x}_u \cdot U_v.)$$

We can solve simultaneous equations to obtain a, b, c and d. For example, multiplying the first equation by $-F$, the second by E and adding produces $-F\ell + Em = -(EG - F^2)b$. Now, $EG - F^2 = |\mathbf{x}_u \times \mathbf{x}_v|^2 \neq 0$, so b is determined. Similarly, we obtain

$$a = \frac{Fm - \ell G}{EG - F^2},$$
$$b = \frac{F\ell - Em}{EG - F^2},$$
$$c = \frac{Fn - Gm}{EG - F^2},$$
$$d = \frac{Fm - En}{EG - F^2}.$$

If we think of these coefficients as a matrix

$$A = \begin{pmatrix} -a & -c \\ -b & -d \end{pmatrix},$$

then, recalling that the *trace* of A and *determinant* of A are given respectively by $\text{tr}(A) = -a - d$ and $\det(A) = ad - bc$, we have

$$\text{tr}(A) = \frac{En + G\ell - 2Fm}{EG - F^2} = 2H,$$

$$\det(A) = \frac{(EG - F^2)(\ell n - m^2)}{(EG - F^2)^2}$$

$$= \frac{\ell n - m^2}{EG - F^2}.$$

This description of mean curvature allows us to write another formula for it which will be useful later.

Proposition 2.5.1.

$$U_v \times \mathbf{x}_u - U_u \times \mathbf{x}_v = 2H\,\mathbf{x}_u \times \mathbf{x}_v.$$

Proof. Using the notation in (2.5.1) for U_u and U_v, we have

$$U_v \times \mathbf{x}_u - U_u \times \mathbf{x}_v = d\,\mathbf{x}_v \times \mathbf{x}_u - a\,\mathbf{x}_u \times \mathbf{x}_v$$

$$= (-a - d)\,\mathbf{x}_u \times \mathbf{x}_v$$

$$= 2H\,\mathbf{x}_u \times \mathbf{x}_v.$$

$\square$

Returning to the matrix A, we see that the equation $\text{tr}(A) = 2H$ gives mean curvature in terms of a linear algebraic invariant of A. It seems reasonable then that other linear algebraic quantities associated to A should also give interesting geometric invariants. In particular, the equation $\det(A) = ad - bc$ should also have some such significance.

Definition 2.5.2. The *Gauss curvature* K is defined to be

$$K = \frac{\ell n - m^2}{EG - F^2}.$$

For an explanation of the geometric significance of the matrix A, see [**Opr97**] for instance. The Gauss curvature, just like the mean curvature, is computed in terms of ℓ, m, n, etc., but unlike mean curvature, it can be calculated in another way in terms of the metric coefficients alone! So, the Gauss curvature is actually intrinsic to the surface, while the mean curvature depends on how the surface sits in space. We now give the formula for K, which obviates the use of U and shows that K depends only on E, F, and G. Although a more general formula exists, we shall restrict ourselves to the case where $F = \mathbf{x}_u \cdot \mathbf{x}_v = 0$. That is, we shall assume that the u- and v-parameter curves always meet at right angles. In this case we state Gauss's *Theorem Egregium*:

Theorem 2.5.3. *The Gauss curvature depends only on the metric E, $F = 0$ and G:*

$$K = -\frac{1}{2\sqrt{EG}} \left(\frac{\partial}{\partial v} \left(\frac{E_v}{\sqrt{EG}} \right) + \frac{\partial}{\partial u} \left(\frac{G_u}{\sqrt{EG}} \right) \right).$$

Here, we have used the notation

$$E_v = \frac{\partial}{\partial v} E = \frac{\partial}{\partial v}(\mathbf{x}_u \cdot \mathbf{x}_u) \quad \text{and} \quad G_u = \frac{\partial}{\partial u} G = \frac{\partial}{\partial u}(\mathbf{x}_v \cdot \mathbf{x}_v).$$

Of course, none of this means anything unless we can compute Gauss and mean curvatures. So far, we have (see (2.2.1)) $\ell = -\mathbf{x}_u \cdot U_u$, etc., and this is rather tedious to compute. We can again use the product rule, however, to make things a bit simpler.

Proposition 2.5.4. *The coefficients of the second fundamental form may be calculated by*

$$\ell = \mathbf{x}_{uu} \cdot U, \quad m = \mathbf{x}_{uv} \cdot U, \quad n = \mathbf{x}_{vv} \cdot U.$$

Proof. $\mathbf{x}_u$ and $\mathbf{x}_v$ are tangent vectors, so $\mathbf{x}_u \cdot U = 0$ and $\mathbf{x}_v \cdot U = 0$. Differentiate both sides of these equations with respect to u and v and use the product rule to obtain

$$\mathbf{x}_{uu} \cdot U + \mathbf{x}_u \cdot U_u = 0, \quad \mathbf{x}_{uv} \cdot U + \mathbf{x}_u \cdot U_v = 0, \quad \mathbf{x}_{vv} \cdot U + \mathbf{x}_v \cdot U_v = 0.$$

By the definitions of ℓ, m and n, we are done. ☐

This result may be used to, rather automatically, calculate Gauss and mean curvatures.

Exercise 2.5.5. Let a surface M be defined by a function of two variables $z = f(x, y)$ with Monge parametrization $(u, v, f(u, v))$. Verify or calculate the following:

(1) $\mathbf{x}_u = (1, 0, f_u)$ and $\mathbf{x}_v = (0, 1, f_v)$.

(2) $\mathbf{x}_{uu} = (0, 0, f_{uu})$ and $\mathbf{x}_{vv} = (0, 0, f_{vv})$.

(3) $\mathbf{x}_u \times \mathbf{x}_v = (-f_u, -f_v, 1)$ so $U = (-f_u, -f_v, 1)/\sqrt{1 + f_u{}^2 + f_v{}^2}$.

(4) Calculate ℓ, n and m.

(5) Using the formula $H = (En + G\ell - 2Fm)/2(EG - F^2)$, show that the condition $H = 0$ reduces to the partial differential equation

$$f_{uu}(1 + f_v^2) + f_{vv}(1 + f_u^2) - 2f_u f_v f_{uv} = 0.$$

This is known as the *minimal surface equation*.

Example 2.5.6 (Surfaces of Revolution). Recall that surfaces of revolution have parametrizations of the form $\mathbf{x}(u, v) = (g(u), h(u) \cos v, h(u) \sin v)$. Also, recall that $g(u)$ is the distance along the axis of revolution and $h(u)$ is the radius of a parallel. We have $\mathbf{x}_u = (g', h' \cos v, h' \sin v)$, $\mathbf{x}_v = (0, -h \sin v, h \cos v)$ and $\mathbf{x}_u \times \mathbf{x}_v = (hh', -g'h \cos v, -g'h \sin v)$. Hence,

$$U = \frac{(h', -g' \cos v, -g' \sin v)}{\sqrt{g'^2 + h'^2}}.$$

Also, the second partial derivatives are $\mathbf{x}_{uu} = (g'', h'' \cos v, h'' \sin v)$, $\mathbf{x}_{uv} = (0, -h' \sin v, h' \cos v)$, and $\mathbf{x}_{vv} = (0, -h \cos v, -h \sin v)$. Hence,

$$E = g'^2 + h'^2, \qquad F = 0, \qquad G = h^2$$

$$l = \frac{(g''h' - h''g')}{\sqrt{g'^2 + h'^2}}, \qquad m = 0, \qquad n = \frac{hg'}{\sqrt{g'^2 + h'^2}}.$$

The mean curvature is given by

$$\begin{aligned}
H &= \frac{G\ell + En - 2Fm}{2(EG - F^2)} \\
&= \frac{h^2 \frac{(g''h' - h''g')}{\sqrt{g'^2 + h'^2}} + (g'^2 + h'^2)\frac{hg'}{\sqrt{g'^2 + h'^2}}}{2(g'^2 + h'^2)h^2} \\
&= \frac{h(g''h' - h''g') + g'(g'^2 + h'^2)}{2h(g'^2 + h'^2)^{3/2}}.
\end{aligned}$$

Finally, the Gauss curvature is computed to be

$$K = \frac{g'(g''h' - h''g')}{h(g'^2 + h'^2)^2}.$$

Exercise 2.5.7. Verify the computations above.

Remark 2.5.8. If the original curve $\alpha(t) = (g(t), h(t), 0)$ has $g'(t) \neq 0$ for all t, then g is strictly increasing. Thus, it is one-to-one and has an inverse function g^{-1} which is differentiable as well. Such an inverse allows us to reparametrize the curve α. Define $f = h \circ g^{-1}$ and get

$$\bar{\alpha}(u) = \alpha \circ g^{-1}(u) = (gg^{-1}(u), hg^{-1}(u), 0)$$
$$= (u, f(u), 0).$$

Thus, our calculations become somewhat easier. For instance, the formula for Gauss curvature becomes $K = -f''/f(1 + f'^2)^2$. To avoid confusion, we will still write

$$\alpha(u) = (u, h(u), 0) \qquad \text{and} \qquad K = -h''/h(1 + h'^2)^2.$$

Exercise 2.5.9. Revolve the catenary $y = c\cosh(x/c)$ around the x-axis to get the catenoid. Show the following (see Example 3.2.1):

$$K = -\frac{1}{c^2 \cosh^4 \frac{u}{c}}, \qquad H = 0.$$

Can you think of another surface of revolution which is minimal? See Theorem 3.2.5.

With a view to giving a short proof of Theorem 2.5.3, we first want to introduce some formulas which will also prove important in Chapter 3 when we link minimal surfaces to complex variable theory. From Lemma 2.1.3, we know that $\{\mathbf{x}_u, \mathbf{x}_v\}$ is a basis for the tangent plane. Because U is normal to the tangent plane, $\{\mathbf{x}_u, \mathbf{x}_v, U\}$ is a basis for $\mathbb{R}^3$. The *acceleration formulas* express the fundamental accelerations $\mathbf{x}_{uu}$, $\mathbf{x}_{uv}$ and $\mathbf{x}_{vv}$ in terms of the basis $\{\mathbf{x}_u, \mathbf{x}_v, U\}$ (at each point of a surface). Again let's assume $F = 0$ for simplicity, and note that this makes $\{\mathbf{x}_u, \mathbf{x}_v, U\}$ an orthogonal basis. Thus, the coefficients may be found by taking dot products. Write

$$\mathbf{x}_{uu} = \Gamma^u_{uu}\mathbf{x}_u + \Gamma^v_{uu}\mathbf{x}_v + \ell U,$$
$$\mathbf{x}_{uv} = \Gamma^u_{uv}\mathbf{x}_u + \Gamma^v_{uv}\mathbf{x}_v + mU,$$
$$(**)\qquad \mathbf{x}_{vv} = \Gamma^u_{vv}\mathbf{x}_u + \Gamma^v_{vv}\mathbf{x}_v + nU.$$

Our job is to find the Γ's. These are just coefficients in a basis expansion, but they are known by the name *Christoffel symbols*, especially in higher dimensional differential geometry and relativity theory. We use what we know about dot products to determine the Γ's:

$$\mathbf{x}_{uu} \cdot \mathbf{x}_u = \Gamma^u_{uu}\mathbf{x}_u \cdot \mathbf{x}_u + 0 + 0$$
$$= \Gamma^u_{uu}E \qquad \text{by definition of } E.$$

We can compute $\mathbf{x}_{uu} \cdot \mathbf{x}_u$ by the product rule and thereby obtain Γ^u_{uu}:

$$E = \mathbf{x}_u \cdot \mathbf{x}_u, \quad \text{so} \quad E_u = \mathbf{x}_{uu} \cdot \mathbf{x}_u + \mathbf{x}_u \cdot \mathbf{x}_{uu} = 2\mathbf{x}_{uu} \cdot \mathbf{x}_u.$$

Thus

$$\mathbf{x}_{uu} \cdot \mathbf{x}_u = \frac{E_u}{2} \quad \text{and} \quad \Gamma^u_{uu} = \frac{E_u}{2E}.$$

Further, $\mathbf{x}_u \cdot \mathbf{x}_v = 0$, so taking the partial with respect to u gives

$$0 = \mathbf{x}_{uu} \cdot \mathbf{x}_v + \mathbf{x}_u \cdot \mathbf{x}_{uv} \quad \text{or} \quad \mathbf{x}_{uu} \cdot \mathbf{x}_v = -\mathbf{x}_u \cdot \mathbf{x}_{uv}.$$

Also, $E = \mathbf{x}_u \cdot \mathbf{x}_u$, so taking the partial with respect to v gives $E_v = 2\mathbf{x}_u \cdot \mathbf{x}_{uv}$ and, consequently, $E_v/2 = \mathbf{x}_u \cdot \mathbf{x}_{uv} = -\mathbf{x}_{uu} \cdot \mathbf{x}_v$. Moreover,

$$\Gamma^v_{uu} = (\mathbf{x}_{uu} \cdot \mathbf{x}_v)/G = -E_v/2G \quad \text{and} \quad \Gamma^u_{uv} = \mathbf{x}_{uv} \cdot \mathbf{x}_u/E = E_v/2E.$$

Continuing on, $G = \mathbf{x}_v \cdot \mathbf{x}_v$, so $G_u/2 = \mathbf{x}_{uv} \cdot \mathbf{x}_v$. Then, since $0 = \mathbf{x}_v \cdot \mathbf{x}_u$, we have

$$-\mathbf{x}_v \cdot \mathbf{x}_{uv} = \mathbf{x}_{vv} \cdot \mathbf{x}_u \quad \text{with} \quad \Gamma^v_{uv} = \mathbf{x}_{uv} \cdot \mathbf{x}_v/G = G_u/2G,$$

$$\Gamma^u_{vv} = \mathbf{x}_{vv} \cdot \mathbf{x}_u/E = -G_u/2E.$$

Finally, $\mathbf{x}_v \cdot \mathbf{x}_v = G$, so $\mathbf{x}_{vv} \cdot \mathbf{x}_v = G_v/2$ and $\Gamma^v_{vv} = \mathbf{x}_{vv} \cdot \mathbf{x}_v/G = G_v/2G$. We end up with the following *acceleration formulas*.

Formula(s) 2.5.10.

$$\mathbf{x}_{uu} = \frac{E_u}{2E}\mathbf{x}_u - \frac{E_v}{2G}\mathbf{x}_v + \ell U,$$

$$\mathbf{x}_{uv} = \frac{E_v}{2E}\mathbf{x}_u + \frac{G_u}{2G}\mathbf{x}_v + mU,$$

$$\mathbf{x}_{vv} = -\frac{G_u}{2E}\mathbf{x}_u + \frac{G_v}{2G}\mathbf{x}_v + nU,$$

$$U_u = -\frac{l}{E}\mathbf{x}_u - \frac{m}{G}\mathbf{x}_v,$$

$$U_v = -\frac{m}{E}\mathbf{x}_u - \frac{n}{G}\mathbf{x}_v.$$

Proof of Theorem 2.5.3. Because mixed partial derivatives are equal no matter the order of differentiation,

$$\mathbf{x}_{uuv} = \mathbf{x}_{uvu}, \quad \text{or} \quad \mathbf{x}_{uuv} - \mathbf{x}_{uvu} = 0.$$

This means that the coefficients of $\mathbf{x}_u$, $\mathbf{x}_v$ and U are *all zero* when $\mathbf{x}_{uuv} - \mathbf{x}_{uvu}$ is written in this basis. Let's concentrate on the $\mathbf{x}_v$-term:

$$\mathbf{x}_{uuv} = \left(\frac{E_u}{2E}\right)_v \mathbf{x}_u + \frac{E_u}{2E}\mathbf{x}_{uv} - \left(\frac{E_v}{2G}\right)_v \mathbf{x}_v - \frac{E_v}{2G}\mathbf{x}_{vv} + \ell_v U + \ell U_v.$$

Now replace $\mathbf{x}_{uv}$, $\mathbf{x}_{vv}$ and U_v by their basis expansions (see Formula(s) 2.5.10) to get,

$$\mathbf{x}_{uuv} = [\ ]\mathbf{x}_u + \left[\frac{E_u G_u}{4EG} - \left(\frac{E_v}{2G}\right)_v - \frac{E_v G_v}{4G^2} - \frac{\ell n}{G}\right]\mathbf{x}_v + [\ ]U$$

$$\mathbf{x}_{uvu} = \left(\frac{E_v}{2E}\right)_u \mathbf{x}_u + \frac{E_v}{2E}\mathbf{x}_{uu} + \left(\frac{G_u}{2G}\right)_u \mathbf{x}_v + \frac{G_u}{2G}\mathbf{x}_{uv} + m_u U + mU_u$$

$$\mathbf{x}_{uvu} = [\ ]\mathbf{x}_u + \left[-\frac{E_v E_v}{4EG} + \left(\frac{G_u}{2G}\right)_u + \frac{G_u G_u}{4G^2} - \frac{m^2}{G}\right]\mathbf{x}_v + [\ ]U.$$

Because the $\mathbf{x}_v$-coefficient of $\mathbf{x}_{uuv} - \mathbf{x}_{uvu}$ is zero, we get

$$0 = \frac{E_u G_u}{4EG} - \left(\frac{E_v}{2G}\right)_v - \frac{E_v G_v}{4G^2} + \frac{E_v E_v}{4EG} - \left(\frac{G_u}{2G}\right)_u - \frac{G_u G_u}{4G^2} - \frac{\ell n - m^2}{G}.$$

Dividing by E, we have

$$\frac{\ell n - m^2}{EG} = \frac{E_u G_u}{4E^2 G} - \frac{1}{E}\left(\frac{E_v}{2G}\right)_v - \frac{E_v G_v}{4EG^2} + \frac{E_v E_v}{4E^2 G} - \frac{1}{E}\left(\frac{G_u}{2G}\right)_u - \frac{G_u G_u}{4EG^2}.$$

Of course, the left-hand side is K (since $F = 0$) and the right-hand side only depends on E and G. Thus, we have a formula for K which does not make explicit use of the normal U. It is now fairly straightforward algebra to show that the right-hand side above is

$$-\frac{1}{2\sqrt{EG}} \left(\frac{\partial}{\partial v} \left(\frac{E_v}{\sqrt{EG}} \right) + \frac{\partial}{\partial u} \left(\frac{G_u}{\sqrt{EG}} \right) \right).$$

$\square$

Chapter 3

The Mathematics of Soap Films

3.1. The Connection

In Chapter 1, we saw that surface tension and the difference in pressure on opposite sides of a soap film combine to determine the 'average' curvature of the soap film. Corollary 2.3.2 emphasized this point by writing the Laplace-Young equation in a form involving the mean curvature H. Specifically, we had

$$p = \sigma \left(\frac{1}{R_1} + \frac{1}{R_2} \right) = 2\sigma H,$$

where σ is surface tension and H is mean curvature. Consider the case of a soap film bounded by a wire, say. Since there is no enclosed volume, the pressure is the same on both sides of the film. Hence, $p = 0$ and (since σ is constant) $H = 0$ as well. Therefore, remembering (see Definition 2.3.3) that a minimal surface is a surface with zero mean curvature at each of its points, we see that

Theorem 3.1.1. *Every soap film is a physical model of a minimal surface.*

We have seen in Theorem 1.4.1 (i.e. the First Principle of Soap Films) that surface tension tries to shrink the surface as much as possible so that it has least surface area among surfaces satisfying certain constraints such as having fixed boundaries or fixed enclosed volumes. In fact, this area-minimizing property implies the vanishing of mean curvature and gives rise to the term *minimal*. Before we see how this works mathematically, let's look at some simple examples.

3.2. The Basics of Minimal Surfaces

Let's begin with two examples of minimal surfaces which we have all seen, the catenoid and helicoid (see § 1.3.10 and § 1.3.9).

Example 3.2.1 (A Catenoid). A catenoid is a surface of revolution generated by a catenary $y(x) = \cosh(x)$ and parametrized by $\mathbf{x}(u,v) = (u, \cosh(u)\cos(v), \cosh(u)\sin(v))$. Let's compute the mean curvature H. First, the tangent vectors are

$$\mathbf{x}_u = (1, \sinh(u)\cos(v), \sinh(u)\sin(v)),$$
$$\mathbf{x}_v = (0, -\cosh(u)\sin(v), \cosh(u)\cos(v)),$$

with $E = \mathbf{x}_u \cdot \mathbf{x}_u = 1 + \sinh^2(u) = \cosh^2(u)$, $F = \mathbf{x}_u \cdot \mathbf{x}_v = 0$ and $G = \mathbf{x}_v \cdot \mathbf{x}_v = \cosh^2(u)$. (Because $E = G$ and $F = 0$, the parametrization is called *isothermal*; and this will be an important idea in studying minimal surfaces.) To get H, we also need the unit normal U and the quantities $\ell = \mathbf{x}_{uu} \cdot U$, $m = \mathbf{x}_{uv} \cdot U$ and $n = \mathbf{x}_{vv} \cdot U$. Of course, we obtain U by taking the cross product of the tangent vectors and dividing by the magnitude of the cross product:

$$\mathbf{x}_u \times \mathbf{x}_v = \det \begin{pmatrix} \mathbf{i} & \mathbf{j} & \mathbf{k} \\ 1 & \sinh(u)\cos(v) & \sinh(u)\sin(v) \\ 0 & -\cosh(u)\sin(v) & \cosh(u)\cos(v) \end{pmatrix}$$

$$= (\sinh(u)\cosh(u), -\cosh(u)\cos(v), -\cosh(u)\sin(v));$$

$$U = (\sinh(u)\cosh(u), -\cosh(u)\cos(v), -\cosh(u)\sin(v))/\cosh^2(u)$$
$$= \left(\frac{\sinh(u)}{\cosh(u)}, -\frac{\cos(v)}{\cosh(u)}, -\frac{\sin(v)}{\cosh(u)} \right).$$

Now, we also have the second partials

$$\mathbf{x}_{uu} = (0, \cosh(u)\cos(v), \cosh(u)\sin(v)),$$
$$\mathbf{x}_{uv} = (0, -\sinh(u)\sin(v), \sinh(u)\cos(v)),$$
$$\mathbf{x}_{vv} = (0, -\cosh(u)\cos(v), -\cosh(u)\sin(v)),$$

with

$$\ell = -\cos^2(v) - \sin^2(v) = -1, \quad m = 0, \quad n = \cos^2(v) + \sin^2(v) = 1.$$

Finally, we get

$$H = \frac{G\ell + En - 2Fm}{2(EG - F^2)} = \frac{\cosh^2(u)(-1) + \cosh^2(u)(1) - 0}{2\cosh^4(u)} = 0.$$

Therefore, the catenoid is a minimal surface, which we saw in § 1.3.10. (Also see § 5.5 and § 5.6.)

Exercise 3.2.2 (A Helicoid). Compute the mean curvature for the helicoid parametrized by

$$\mathbf{x}(u, v) = (v\cos(u), v\sin(u), u).$$

(Also see § 5.5 for automated calculation and § 1.3.9 for the relevant soap film.)

Now let's look at one of our other standard situations: a surface obtained as the graph of a function of two variables. Let $z = f(x, y)$ be a function of two variables, and take a Monge parametrization for its graph: $\mathbf{x}(u, v) = (u, v, f(u, v))$. We have

$$\mathbf{x}_u = (1, 0, f_u), \qquad \mathbf{x}_{uu} = (0, 0, f_{uu}),$$
$$\mathbf{x}_v = (0, 1, f_v), \qquad \mathbf{x}_{uv} = (0, 0, f_{uv}),$$
$$\mathbf{x}_{vv} = (0, 0, f_{vv}),$$

$$\mathbf{x}_u \times \mathbf{x}_v = (-f_u, -f_v, 1), \qquad U = \frac{(-f_u, -f_v, 1)}{\sqrt{1 + f_u^2 + f_v^2}},$$

$$E = 1 + f_u^2, \qquad F = f_u f_v, \qquad G = 1 + f_v^2,$$

$$\ell = \frac{f_{uu}}{\sqrt{1 + f_u^2 + f_v^2}}, \quad m = \frac{f_{uv}}{\sqrt{1 + f_u^2 + f_v^2}}, \quad n = \frac{f_{vv}}{\sqrt{1 + f_u^2 + f_v^2}},$$

$$K = \frac{f_{uu}f_{vv} - f_{uv}^2}{(1 + f_u^2 + f_v^2)^2},$$

$$H = \frac{(1 + f_v^2)f_{uu} + (1 + f_u^2)f_{vv} - 2f_u f_v f_{uv}}{2(1 + f_u^2 + f_v^2)^{\frac{3}{2}}}.$$

The expression for H and Definition 2.3.3 immediately show the following for M given by the graph of $z = f(x, y)$.

Proposition 3.2.3. *M is minimal if and only if*

$$f_{uu}(1 + f_v^2) - 2f_u f_v f_{uv} + f_{vv}(1 + f_u^2) = 0.$$

This equation is called the *minimal surface equation*, and, in general, it is not solvable by simple means. However, we can sometimes hypothesize algebraic or geometric requirements about the function f which allow us to solve the minimal surface equation and thereby determine certain types of minimal surfaces. Here is the most famous instance of this.

Example 3.2.4 (Scherk's First Surface). Suppose we require the algebraic condition that the function f be separately dependent on the variables x and y. This seems like a natural algebraic condition to put on a function. Also, it is a standard trick (multiplicatively) in solving partial differential equations, so it doesn't come out of thin air. So, write $f(x, y) = g(x) + h(y)$. The minimal surface equation then becomes

$$g''(x)(1 + h'^2(y)) + h''(y)(1 + g'^2(x)) = 0,$$

where the primes refer to derivatives with respect to the appropriate variables. But this is an example of a *separable* ordinary differential equation, and *these are very solvable*. Put all x's on one side and all y's on the other to obtain

$$-\frac{1 + g'^2(x)}{g''(x)} = \frac{1 + h'^2(y)}{h''(y)}.$$

Now how can this equality possibly be true when one side is a function of x alone and the other side is a function of y alone? If we keep y fixed, the right side stays fixed, so varying x has no effect. Similarly, varying y has no effect if x is kept fixed. This means that the only way the equality can hold is if both sides are equal to the same constant, say c. Let's look at the x side then. We have

$$-\frac{1 + g'^2(x)}{g''(x)} = c$$

$$1 + g'^2(x) = -c\,g''(x),$$

and this equation does not involve g itself. So, now we can substitute $\phi = g'$ to get $1 + \phi^2 = -c\,\phi'$. Upon separating and integrating, we obtain

$$\int \frac{-c\,d\phi}{1 + \phi^2} = \int dx.$$

A simple trigonometric substitution $\phi = \tan(\theta)$ and integration produces $x = -c\arctan(\phi)$ (where we have suppressed the constant of integration). We then get $\phi = -\tan(x/c)$, and replacing ϕ by g' and integrating again gives $g(x) = c\ln(\cos(x/c))$. The same calculation holds for the y-side (without the minus sign, of course) to give $h(x) = -c\ln(\cos(y/c))$, so we finally obtain an expression for f:

$$f(x, y) = c\ln(\cos(x/c)) - c\ln(\cos(y/c)) = c\ln\left(\frac{\cos(x/c)}{\cos(y/c)}\right).$$

The surface $z = f(x, y)$ is called *Scherk's (first) minimal surface*. Note that Scherk's surface is only defined for $\frac{\cos(x/c)}{\cos(y/c)} > 0$. For example, letting $c = 1$ for covenience, a piece of Scherk's surface is defined over the square $-\pi/2 < x < \pi/2$, $-\pi/2 < y < \pi/2$. The pieces of Scherk's surface fit together according to the Schwarz reflection principles (see Theorem 3.9.9). Surprisingly, only the catenoid (see Theorem 3.2.5) and helicoid were known to be minimal in the 1700's. Scherk's surface was the next example of a minimal surface, and it was discovered by Scherk in 1835. For a picture of Scherk's first surface see § 5.7.

We just saw how an algebraic requirement turns the minimal surface equation into a pair of simple ordinary differential equations. Now let's impose some geometric requirements. For example, what if we require that a minimal surface be a surface of revolution? Does this imposition of circular symmetry somehow geometrically constrain the minimal surface equation to make it solvable? Here is the answer.

Theorem 3.2.5. *If a surface of revolution* M *is minimal, then* M *is contained in either a plane or a catenoid.*

Proof. For simplicity, we will take the special case of a parametrization $\mathbf{x}(u, v) = (u,\ h(u)\cos v, h(u)\sin v)$ for M. Then, using Example 2.5.6 with $g(u) = u$, we have

$$H = \frac{1}{2}\left(\frac{-hh'' + 1 + h'^2}{h(1 + h'^2)^{3/2}}\right).$$

The surface M is minimal, so $H = 0$ and, thus, $hh'' = 1 + h'^2$. Let $w = h'$. Then

$$h'' = w' = \frac{dw}{dh}\frac{dh}{du} = \frac{dw}{dh}w.$$

Here we consider w as a function of h on an interval with $h'(u) \neq 0$. We can do this because, on such an interval, h has an inverse function f with $u = f(h)$. Then, taking a derivative and applying the chain rule gives

$$1 = \frac{df}{dh}\frac{dh}{du} = \frac{df}{dh}w \quad \text{or} \quad w = \frac{1}{\frac{df}{dh}},$$

so that w is a function of h since f and df/dh are. Thus, $hh'' = 1 + h'^2$ implies $hw(dw/dh) = 1 + w^2$; or $\frac{w}{1+w^2}dw = \frac{1}{h}dh$. We can integrate both sides to get

$$\ln(\sqrt{1 + w^2}) = \ln h + c$$
$$\sqrt{1 + w^2} = ch$$
$$w = \sqrt{c^2h^2 - 1}.$$

Now, $w = \frac{dh}{du}$, so $\frac{1}{\sqrt{c^2h^2-1}}dh = du$, and integration yields (with $ch = \cosh s$)

$$\frac{1}{c}s = u + D$$

$$\frac{1}{c}\cosh^{-1}(ch) = u + D$$

$$ch = \cosh(cu + k)$$

$$h(u) = \frac{1}{c}\cosh(cu + k).$$

Therefore, M is part of a catenoid. □

Exercise 3.2.6. In the proof above we assumed that $h'(u) \neq 0$. What happens if $h'(u) = 0$? Explain.

Exercise 3.2.7. What if we don't make the simplifying assumption that $g(u) = u$? Discuss this by considering three cases: (1) $g'(u) = 0$ on an interval, so the surface is part of a plane (why?); (2) $g'(u) \neq 0$ on an interval, so the inverse function theorem (look it up) reduces this to the case $g(u) = u$; (3) $g'(u) = 0$ at isolated points. Show this can't happen by comparing the solution surrounding the isolated point having $g'(u) = 0$ with what happens at the point itself. Remember, the profile curve $\alpha(u) = (g(u), h(u), 0)$ has slope $h'(u)/g'(u)$.

Naturally, we can ask similar questions by imposing other geometric hypotheses. In order to mention another result in this direction, recall that a surface is *ruled* if it has a parametrization

$$\mathbf{x}(u, v) = \beta(u) + v\delta(u)$$

where β and δ are curves. That is, the entire surface is covered by this one parametrization, which consists of lines emanating from a curve $\beta(u)$ going in the direction $\delta(u)$. The curve $\beta(u)$ is called the *directrix* of the surface, and a line having $\delta(u)$ as direction vector is called a *ruling*. For instance, cones are given by $\mathbf{x}(u, v) = p + v\delta(u)$, where p is a fixed point, and cylinders are given by $\mathbf{x}(u, v) = \beta(u) + vq$, where q is a fixed direction vector. The helicoid is ruled as well, as can be seen by its parametrization

$$\mathbf{x}(u, v) = (av\cos(u), av\sin(u), bu) = (0, 0, bu) + v(a\cos(u), a\sin(u), 0).$$

Ruled surfaces are useful in such areas of engineering as gear design and nuclear cooling tower design, so we might ask when they can be achieved by soap films also. The answer is, not too often. In fact, there is

Theorem 3.2.8 (Catalan's Theorem). *Any ruled minimal surface in* $\mathbb{R}^3$ *is part of a plane or a helicoid.*

The proof of this result requires more curve theory than we want to get into, so we refer to [**BC86**] or [**Opr97**] for example. Theorem 3.2.5 and Theorem 3.2.8 show us that the algebra of calculating mean curvature to be zero often is mirrored in geometric reality.

Remark 3.2.9. The parametrization

$$\mathbf{x}(u,v) = (x^1(u,v), x^2(u,v), x^3(u,v)),$$

with

$$x^1(u,v) = \cos(t)\sinh(v)\sin(u) + \sin(t)\cosh(v)\cos(u),$$
$$x^2(u,v) = -\cos(t)\sinh(v)\cos(u) + \sin(t)\cosh(v)\sin(u),$$
$$x^3(u,v) = u\cos(t) + v\sin(t),$$

gives the catenoid for $t = \pi/2$ and (an isometric copy of) the helicoid $\mathbf{x}(u,v) = (v\cos u, v\sin u, u)$ for $t = 0$. The helicoid and catenoid are *associated* in a sense to be made precise when we talk about the Weierstrass-Enneper representation later. At any rate, the association of the catenoid and the helicoid makes Theorem 3.2.8 seem a natural companion to Theorem 3.2.5.

Example 3.2.10 (Some Famous Minimal Surfaces). The following are some minimal surfaces which are well known. See § 5.7 for Maple plots. Once we know something about the Weierstrass-Enneper representation of minimal surfaces, their parametrizations won't seem so mysterious. The reader is invited to check that $H = 0$ for each.

(1) *Enneper's Surface.*

$$\mathbf{x}(u,v) = (u - u^3/3 + uv^2, v - v^3/3 + vu^2, u^2 - v^2).$$

(2) *Henneberg's Surface.* $\mathbf{x}(u,v) = (x^1(u,v), x^2(u,v), x^3(u,v))$,
where

$$x^1(u,v) = 2\sinh(u)\cos(v) - \frac{2}{3}\sinh(3u)\cos(3v),$$

$$x^2(u,v) = 2\sinh(u)\sin(v) + \frac{2}{3}\sinh(3u)\sin(3v),$$

$$x^3(u,v) = 2\cosh(2u)\cos(2v).$$

(3) *Catalan's Surface.*

$$\mathbf{x}(u,v) = (u - \sin(u)\cosh(v), 1 - \cos(u)\cosh(v), 4\sin(u/2)\sinh(v/2)).$$

(4) *Scherk's Fifth Surface.* This surface is often written in non-parametric form $\sin z = \sinh x \sinh y$. Parametrically,

$$\mathbf{x}(u,v) = (\text{arcsinh}(u), \text{arcsinh}(v), \arcsin(uv)).$$

3.3. Area Minimization and Soap Films

As his experiments with soap films progressed, Plateau became convinced that any type of boundary would produce an enclosed soap film. Conceivably, however, there could be a boundary so complicated that the required twists and turns of the film would cause it to break, so Plateau set forth the following problem, now known as *Plateau's problem*: given a curve C, find a minimal surface M having C as boundary. As we shall see below, *least-area* surfaces are minimal, and it was eventually realized that hopes of finding a general solution for Plateau's problem rested on finding least-area surfaces. Thus, another version of Plateau's problem is to find a least-area surface having C as boundary. Of course, by Theorem 1.4.1, this takes us right back to soap films. Even the *existence* of area-minimizing surfaces is not automatic, however. It was only early in this century that Plateau's problem was solved by J. Douglas and T. Rado (see [**Dou31**] and [**Rad71**]). They proved

Theorem 3.3.1. *There exists a least-area disk-like minimal surface spanning any given Jordan curve.*

(A minimal surface is *disk-like* if its parameter domain is the unit disk $D = \{(u, v)|u^2 + v^2 \leq 1\}$ and the boundary circle of the disk maps to the given Jordan curve.)

Instead of looking at the difficult question of existence, let's ask what a surface M having least area among surfaces with specified boundary C implies about the surface. In this way, we obtain a necessary condition for M's existence. Of course, if we are talking about surface area, we must have an understanding of how to compute it. In the plane, we do ordinary double integration to calculate area. To deal with curved surfaces, we use the defining parametrizations to transport all calculus computations done in the plane to the given surface. In the plane, we calculate the area of a parallelogram by finding the length of the cross product of the two vectors on its sides. For a surface, we can imagine tiny pieces of area approximated by tiny parallelograms. Since the approximating parallelogram is very small, we may think of it as spanned by tangent vectors multiplied by the changes in the parameters,

$$\mathbf{x}_u \, \Delta u \quad \text{and} \quad \mathbf{x}_v \, \Delta v.$$

Hence, the area is given by $|\mathbf{x}_u \times \mathbf{x}_v| \Delta u \, \Delta v$. This quantity then approximates the area of that tiny piece of the surface. Of course this is what we usually do in calculus: approximate a small piece and then add up continuously (i.e. integrate) over the entire region. So, let $\mathbf{x}(u, v)$ be a parametrization for a surface M and make the

Definition 3.3.2. The area of the parametrization $\mathbf{x}$, denoted $A_{\mathbf{x}}$, is

$$A_{\mathbf{x}} = \iint |\mathbf{x}_u \times \mathbf{x}_v| \, du \, dv,$$

where the limits of integration are the defining limits of the parametrization.

Furthermore, any compact oriented surface may be cut up into a finite number of parametrizations $\mathbf{x}$ (which, in fact may be taken to be oriented triangles) meeting only along their boundary curves with the opposite orientation. Therefore, we may calculate the whole surface area of a compact oriented surface M as the sum of the areas of the individual pieces. Although this seems straightforward, we are truly using the power of calculus here. Surfaces unlike ours which have sharp corners and other places where derivatives don't exist can present problems when we look at area questions. The modern subject of *geometric measure theory* [**Mor88**] is designed to overcome many of these difficulties.

Example 3.3.3. Let $z = f(x, y)$ be a function of two variables. Then we have a Monge parametrization

$$\mathbf{x}(u, v) = (u, v, f(u, v)),$$

and we can calculate $\mathbf{x}_u$ and $\mathbf{x}_v$ to be

$$\mathbf{x}_u = (1, 0, f_u) \qquad \mathbf{x}_v = (0, 1, f_v)$$

with

$$|\mathbf{x}_u \times \mathbf{x}_v| = |(-f_u, -f_v, 1)| = \sqrt{1 + f_u^2 + f_v^2}.$$

Hence, $\text{Area}_{\mathbf{x}} = \iint \sqrt{1 + f_u^2 + f_v^2} \, du \, dv$. We shall use this in Example 4.4.8.

One convenient way to calculate the area enclosed by closed curves in the plane is by using Green's theorem. Suppose P and Q are two real valued (smooth) functions of two variables x and y defined on a simply connected region of the plane. Then Green's theorem says that

$$\int_y \int_x \frac{\partial P}{\partial x} + \frac{\partial Q}{\partial y} \, dx \, dy = \int_C -Q \, dx + P \, dy,$$

where the right-hand side is the line integral around the boundary C of the region enclosed by C. Because we pull all integrals back to the plane for computation, we will find Green's theorem particularly useful.

Exercise 3.3.4. Apply Green's theorem to show that the area inside a closed curve C is given by

$$\frac{1}{2} \oint_C -y \, dx + x \, dy.$$

Now suppose M is the graph of a function of two variables $z = f(x, y)$ and is a surface of least area with boundary C. Consider the nearby surfaces, which look like slightly deformed versions of M,

$$M^t \colon z^t(x, y) = f(x, y) + tg(x, y).$$

Here, g is a function on the domain of f with $g|_{\tilde{C}} = 0$ (where $\tilde{C}$ is the boundary of the domain of f and $f(\tilde{C}) = C$). The perturbation $tg(x, y)$ then has the effect of moving points of M a small bit and leaving C fixed. A Monge parametrization for M^t is given by

$$\mathbf{x}^t(u, v) = (u, v, f(u, v) + tg(u, v)).$$

So let's now find a description of M's area. First, we have

$$|\mathbf{x}_u^t \times \mathbf{x}_v^t| = \sqrt{1 + f_u^2 + f_v^2 + 2t(f_u g_u + f_v g_v) + t^2(g_u{}^2 + g_v{}^2)}.$$

By the definition of area, we see that the area of M^t is

$$A(t) = \int_v \int_u \sqrt{1 + f_u^2 + f_v^2 + 2t(f_u g_u + f_v g_v) + t^2(g_u{}^2 + g_v{}^2)} \, du \, dv.$$

Now that that is done, what do we know? Well, M is assumed to have least area, so the area function $A(t)$ must have a minimum at $t = 0$ (since $t = 0$ gives us M). So let's take a derivative of $A(t)$ and impose the condition that $A'(0) = 0$. We can take the derivative with respect to t inside the integral (essentially because t has nothing to do with the parameters u and v, but also see Lemma 4.3.1) and use the chain rule to get

$$A'(t) = \int_v \int_u \frac{f_u g_u + f_v g_v + t(g_u{}^2 + g_v{}^2)}{\sqrt{1 + f_u^2 + f_v^2 + 2t(f_u g_u + f_v g_v) + t^2(g_u{}^2 + g_v{}^2)}} \, du \, dv.$$

Again, we assumed $z = z_0$ was a minimum, so $A'(0) = 0$. Therefore, setting $t = 0$ in the equation above, we get

$$\int_v \int_u \frac{f_u g_u + f_v g_v}{\sqrt{1 + f_u^2 + f_v^2}} \, du \, dv = 0.$$

Now, let

$$P = \frac{f_u g}{\sqrt{1 + f_u^2 + f_v^2}}, \qquad Q = \frac{f_v g}{\sqrt{1 + f_u^2 + f_v^2}}.$$

Exercise 3.3.5. Compute $\frac{\partial P}{\partial u}$ and $\frac{\partial Q}{\partial v}$ and apply Green's theorem.

We then get

$$\int_v \int_u \frac{f_u g_u + f_v g_v}{\sqrt{1 + f_u^2 + f_v^2}} \, du \, dv$$

$$+ \int_v \int_u \frac{g[f_{uu}(1 + f_v^2) + f_{vv}(1 + f_u^2) - 2f_u f_v f_{uv}]}{(1 + f_u^2 + f_v^2)^{\frac{3}{2}}} \, du \, dv$$

$$= \int_{\tilde{C}} \frac{f_u g \, dv}{\sqrt{1 + f_u^2 + f_v^2}} - \frac{f_v g \, du}{\sqrt{1 + f_u^2 + f_v^2}}$$

$$= 0$$

since $g|_{\tilde{C}} = 0$. Of course the first integral is zero as well, so we end up with

$$\int_v \int_u \frac{g[f_{uu}(1 + f_v^2) + f_{vv}(1 + f_u^2) - 2f_u f_v f_{uv}]}{(1 + f_u^2 + f_v^2)^{\frac{3}{2}}} \, du \, dv = 0.$$

Since this is true for all such g, we must have (see Exercise 4.3.2)

$$f_{uu}(1 + f_v^2) + f_{vv}(1 + f_u^2) - 2f_u f_v f_{uv} = 0.$$

This is the minimal surface equation. Therefore, by the calculation in Proposition 3.2.3, we have $H = 0$, and we have shown the following *necessary* condition for a surface to be area-minimizing.

Theorem 3.3.6. *If M is area-minimizing, then M is minimal.*

Theorem 1.4.1 tells us that soap films tend to minimize area, so this result is another confirmation that the study of soap films is a study of minimal surfaces. The whole method leading to Theorem 3.3.6 is an example of a basic technique of the calculus of variations, and we shall see it again in Chapter 4. In particular, Example 4.4.8 applies the technique directly to the usual surface area integral of calculus for a function of two variables. Finally, § 5.8 and § 3.10 show that *not all* minimal surfaces are area-minimizing, so Theorem 3.3.6 is truly only a necessary condition.

Exercise 3.3.7. This exercise explores the area-minimizing existence question for the catenoid. See § 5.6 for a Maple analysis. A *catenary* is given by $y = a \cosh(x/a)$. Determine the parameter a for the catenary which passes through the points $(-.6, 1)$ and $(.6, 1)$. Compute the surface area of the catenoid generated by revolving the catenary about the x-axis. Show that this surface area is greater than the surface area of two disks of radius 1. Hence, if the original boundary curve consisted of two circles perpendicular to the x-axis centered at $(-.6, 1)$ and $(.6, 1)$, then the catenoid would be a minimal, but non-area-minimizing, surface spanning the boundary.

In fact, more can be shown. Let x_0, $-x_0$ be the points on the x-axis which are centers of the given circles of radius y_0.

(1) If $x_0/y_0 > .528$ (approximately), then the two disks give an absolute minimum for surface area. This is the so-called Goldschmidt discontinuous solution.

(2) If $x_0/y_0 < .528$ (approximately), then a catenoid is the absolute minimum and the Goldschmidt solution is a local minimum.

(3) If $.528 < x_0/y_0 < .663$ (approximately), then the catenoid is only a local minimum.

(4) If $x_0/y_0 > .663$ (approximately), then there is no catenoid joining the points.

This can be seen using soap films by forming a catenoid between two rings and slowly pulling the rings apart (see § 1.3.10). Measure x_0/y_0, and you will see that the catenoid spontaneously jumps to the two-disk solution at some point. How does this jibe with the bounds above? For an informal discussion of this problem, see [**Ise92**]. For a formal discussion, see [**Bli46**].

3.4. Isothermal Parameters

What we have done so far might fairly be called the basics of minimal surface theory. We now have an idea of what minimal surfaces are and why their study is relevant to understanding soap films. Nevertheless, we have only used the tools of differential geometry and differential equations to explore certain aspects of minimal surfaces. In this section we begin to introduce the power of complex analysis into the mix of ideas surrounding soap films. The key to introducing complex analysis into minimal surface theory is the existence of isothermal parameters on a minimal surface. A parametrization $\mathbf{x}(u, v)$ is *isothermal* if $E = \mathbf{x}_u \cdot \mathbf{x}_u = \mathbf{x}_v \cdot \mathbf{x}_v = G$ and $F = 0$. This innocuous-looking assumption about the metric will lead to wonderful consequences. In fact, while isothermal coordinates exist on all surfaces, the proof of this is much harder than the one given below for minimal surfaces.

Theorem 3.4.1. *Isothermal coordinates exist on any minimal surface $M \subseteq \mathbb{R}^3$.*

Proof ([Oss86]). Fix a point $m \in M$. Choose a coordinate system for $\mathbb{R}^3$ so that m is the origin, the tangent plane to M, $T_m M$, is the xy-plane, and near m, M is the graph of a function $z = f(x, y)$. Furthermore, the quotient and chain rules give

$$\left(\frac{1 + f_x^2}{w}\right)_y - \left(\frac{f_x f_y}{w}\right)_x$$

$$= -\frac{f_y}{w}\left[f_{xx}(1 + f_y^2) - 2f_x f_y f_{xy} + f_{yy}(1 + f_x^2)\right],$$

$$\left(\frac{1 + f_y^2}{w}\right)_x - \left(\frac{f_x f_y}{w}\right)_y$$

$$= -\frac{f_x}{w}\left[f_{xx}(1 + f_y^2) - 2f_x f_y f_{xy} + f_{yy}(1 + f_x^2)\right],$$

where $w = \sqrt{1 + f_x^2 + f_y^2}$. Let $p = f_x$, $q = f_y$ with $w^2 = 1 + p^2 + q^2$. Because M is minimal, f satisfies the minimal surface equation

$$f_{xx}(1 + f_y^2) - 2f_x f_y f_{xy} + f_{yy}(1 + f_x^2) = 0,$$

so we have $(\frac{1+p^2}{w})_y - (\frac{pq}{w})_x = 0$ and $(\frac{1+q^2}{w})_x - (\frac{pq}{w})_y = 0$. Define two vector fields in the xy-plane by

$$V = \left(\frac{1 + p^2}{w}, \frac{pq}{w}\right) \quad \text{and} \quad W = \left(\frac{pq}{w}, \frac{1 + q^2}{w}\right)$$

and apply Green's theorem to any closed curve C contained in a simply connected region $\mathcal{R}$ to obtain

$$\int_C V \cdot dr = \iint_{\mathcal{R}} \left(\frac{pq}{w}\right)_x - \left(\frac{1 + p^2}{w}\right)_y \, dx \, dy = 0,$$

$$\int_C W \cdot dr = \iint_{\mathcal{R}} \left(\frac{1+q^2}{w}\right)_x - \left(\frac{pq}{w}\right)_y dx\, dy = 0,$$

where $dr = (dx, dy)$. Since the line integrals are zero for all closed curves in $\mathcal{R}$, V and W must have potential functions (see [**MT88**]). That is, there exist μ and ρ with $\text{grad}(\mu) = V$ and $\text{grad}(\rho) = W$. Considered coordinatewise, these equations imply $\mu_x = \frac{1+p^2}{w}, \mu_y = \frac{pq}{w}$ and $\rho_x = \frac{pq}{w}, \rho_y = \frac{1+q^2}{w}$. Define a mapping $T: \mathcal{R} \to \mathbb{R}^2$ by

$$T(x, y) = (x + \mu(x, y), y + \rho(x, y)).$$

The Jacobian matrix of this mapping is then

$$J(T) = \begin{bmatrix} 1 + \mu_x & \mu_y \\ \rho_x & 1 + \rho_y \end{bmatrix} = \begin{bmatrix} 1 + \frac{1+p^2}{w} & \frac{pq}{w} \\ \frac{pq}{w} & 1 + \frac{1+q^2}{w} \end{bmatrix},$$

and we calculate the determinant to be $\det(J(T)) = \frac{(1+w)^2}{w} > 0$. The Inverse Function Theorem then says that, near $m = (0, 0)$, there is a smooth inverse function $T^{-1}(u, v) = (x, y)$ with

$$J(T^{-1}) = J(T)^{-1}$$

$$= \frac{1}{\det J(T)} \begin{bmatrix} 1 + \frac{1+q^2}{w} & -\frac{pq}{w} \\ -\frac{pq}{w} & 1 + \frac{1+p^2}{w} \end{bmatrix}$$

$$= \frac{1}{(1+w)^2} \begin{bmatrix} w + 1 + q^2 & -pq \\ -pq & w + 1 + p^2 \end{bmatrix}.$$

Of course, for $(x, y) = T^{-1}(u, v)$, the last matrix is just

$$\begin{bmatrix} x_u & x_v \\ y_u & y_v \end{bmatrix}$$

by the definition of the Jacobian. We will put these calculations to use in showing that the parametrization (in the uv coördinates described above)

$$\mathbf{x}(u, v) \overset{\text{def}}{=} (x(u, v), y(u, v), f(x(u, v), y(u, v)))$$

is isothermal. First we calculate

$$\mathbf{x}_u = \left(\frac{w + 1 + q^2}{(1+w)^2}, \frac{-pq}{(1+w)^2}, p\left(\frac{w + 1 + q^2}{(1+w)^2}\right) + q\left(\frac{-pq}{(1+w)^2}\right)\right)$$

and

$$E = \mathbf{x}_u \cdot \mathbf{x}_u$$
$$= \frac{1}{(1+w)^4} \left[(w+1+q^2)^2 + p^2q^2 + p^2(w+1+q^2)^2 \right.$$
$$\left. - 2p^2q^2(w+1+q^2) + p^2q^4 \right]$$
$$= \frac{1}{(1+w)^4} \left[(1+w)^2(1+q^2+p^2) \right]$$
$$= \frac{w^2}{(1+w)^2}.$$

Exercise 3.4.2. Show that

$$\mathbf{x}_v = \left(\frac{-pq}{(1+w)^2}, \frac{w+1+p^2}{(1+w)^2}, p\left(\frac{-pq}{(1+w)^2} \right) + q\left(\frac{w+1+p^2}{(1+w)^2} \right) \right)$$

and $G = \mathbf{x}_v \cdot \mathbf{x}_v = E$, $F = 0$.

Hence, the parametrization $\mathbf{x}(u,v)$ is isothermal. □

Exercise 3.4.3. If M is a surface with isothermal parametrization $\mathbf{x}(u,v)$, show that the formula for mean curvature reduces to $H = \frac{\ell+n}{2E}$. Hence, for M minimal, $\ell = -n$. Recall that this was true for the catenoid (see Example 3.2.1).

3.5. Harmonic Functions and Minimal Surfaces

When a minimal surface is parametrized by an isothermal parametrization $\mathbf{x}(u,v)$, there is a close tie between the Laplace operator $\Delta\mathbf{x} = \mathbf{x}_{uu} + \mathbf{x}_{vv}$ and mean curvature. To see this, we need the acceleration formulas, Formula(s) 2.5.10 of Chapter 2.

Theorem 3.5.1. *If the parametrization* $\mathbf{x}$ *is isothermal, then* $\Delta\mathbf{x} \overset{\text{def}}{=} \mathbf{x}_{uu} + \mathbf{x}_{vv} = (2\,E\,H)\,U$.

Proof.

$$\mathbf{x}_{uu} + \mathbf{x}_{vv} = \left(\frac{E_u}{2E}\mathbf{x}_u - \frac{E_v}{2G}\mathbf{x}_v + l\,U\right) + \left(-\frac{G_u}{2E}\mathbf{x}_u + \frac{G_v}{2G}\mathbf{x}_v + n\,U\right)$$

$$= \frac{E_u}{2E}\mathbf{x}_u - \frac{E_v}{2E}\mathbf{x}_v + l\,U - \frac{E_u}{2E}\mathbf{x}_u + \frac{E_v}{2E}\mathbf{x}_v + n\,U$$

$$= (l+n)\,U$$

$$= 2\,E\left(\frac{l+n}{2E}\right)U.$$

By examining the formula for mean curvature when $E = G$ and $F = 0$, we see that

$$H = \frac{El + En}{2E^2} = \frac{l+n}{2E}.$$

Therefore,

$$\mathbf{x}_{uu} + \mathbf{x}_{vv} = (2\,E\,H)\,U.$$

$\square$

Corollary 3.5.2. *A surface M with an isothermal parametrization $\mathbf{x}(u,v) = (x^1(u,v), x^2(u,v), x^3(u,v))$ is minimal if and only if x^1, x^2 and x^3 are harmonic functions.*

Proof. If M is minimal, then $H = 0$ and, by the previous theorem, $\mathbf{x}_{uu} + \mathbf{x}_{vv} = 0$. Therefore, the coordinate functions of $\mathbf{x}$ are harmonic. On the other hand, suppose x^1, x^2, and x^3 are harmonic functions. Then $\mathbf{x}_{uu} + \mathbf{x}_{vv} = 0$ and, by Theorem 3.5.1, $(2\,E\,H)\,U = 0$. Therefore, since U is the unit normal and $E = \mathbf{x}_u \cdot \mathbf{x}_u \neq 0$, we have $H = 0$, and M is minimal. $\square$

This result links the geometry of minimal surface theory to complex analysis because, as we saw in § 2.4, harmonic functions, together with their harmonic conjugate functions, produce holomorphic functions. Corollary 3.5.2 then naturally leads us into the world of complex variables through the closely aligned subject of harmonic functions. Before we go there, however, let's prove one geometric fact about minimal surfaces using harmonic functions themselves. Recall that a set in $\mathbb{R}^3$ is compact if and only if it is closed and bounded, and that any continuous function on a compact set attains its maximum and minimum values on the set.

Theorem 3.5.3. *If M is a minimal surface without boundary, then M cannot be compact.*

Proof. Give M an isothermal parametrization. By Corollary 3.5.2, the coordinate functions of the parametrization are harmonic functions. If M were compact, then each coordinate function would attain its maximum and minimum. By Theorem 2.4.11, this must happen on the boundary of M. Since M has no boundary, this contradicts compactness. $\square$

3.6. The Weierstrass-Enneper Representations

Finally we come to the heart-and-soul of minimal surface theory: the *Weierstrass-Enneper representation*. What makes this approach so useful and interesting is that it will allow us to create minimal surfaces by just choosing holomorphic functions. What could be simpler?

Let M be a minimal surface described by an isothermal parametrization $\mathbf{x}(u, v)$. Let $z = u + iv$ denote the corresponding complex coordinate, and recall that

$$\frac{\partial}{\partial z} = \frac{1}{2}\left(\frac{\partial}{\partial u} - i\frac{\partial}{\partial v}\right), \qquad \frac{\partial}{\partial \bar{z}} = \frac{1}{2}\left(\frac{\partial}{\partial u} + i\frac{\partial}{\partial v}\right).$$

Since $u = \frac{z+\bar{z}}{2}$ and $v = \frac{-i(z-\bar{z})}{2}$, we may write

$$\mathbf{x}(z, \bar{z}) = \left(x^1(z, \bar{z}), x^2(z, \bar{z}), x^3(z, \bar{z})\right).$$

We regard $x^i(z, \bar{z})$ as a complex valued function which happens to take real values, and we have $\frac{\partial x^i}{\partial z} = \frac{1}{2}(x_u^i - ix_v^i)$ by the definition of $\partial/\partial z$. Define

$$\phi \stackrel{\text{def}}{=} \frac{\partial \mathbf{x}}{\partial z} = (x_z^1, x_z^2, x_z^3).$$

Let's examine ϕ a bit more closely. We shall use the following notation: $(\phi)^2 = (x_z^1)^2 + (x_z^2)^2 + (x_z^3)^2$ and $|\phi|^2 = |x_z^1|^2 + |x_z^2|^2 + |x_z^3|^2$, where $|z| = \sqrt{u^2 + v^2}$ is the modulus of z. First, note that $(x_z^i)^2 =$

$(1/4)((x_u^i)^2 - (x_v^i)^2 - 2ix_u^i x_v^i)$. Therefore,

$$(\phi)^2 = \frac{1}{4}\left(\sum_{j=1}^{3}(x_u^j)^2 - \sum_{j=1}^{3}(x_v^j)^2 - 2i\sum_{j=1}^{3} x_u^j x_v^j\right)$$

$$= \frac{1}{4}(|\mathbf{x}_u|^2 - |\mathbf{x}_v|^2 - 2i\mathbf{x}_u \cdot \mathbf{x}_v)$$

$$= \frac{1}{4}(E - G - 2iF)$$

$$= 0,$$

since $\mathbf{x}(u,v)$ is isothermal. By comparing real and imaginary parts, we see that the converse is true as well. Namely, if $(\phi)^2 = 0$, then the parametrization must be isothermal.

Exercise 3.6.1. Show that $|\phi|^2 = \frac{E}{2} \neq 0$ by verifying the following calculation:

$$|\phi|^2 = \frac{1}{4}\left(\sum_{j=1}^{3}(x_u^j)^2 + \sum_{j=1}^{3}(x_v^j)^2\right)$$

$$= \frac{1}{4}(|\mathbf{x}_u|^2 + |\mathbf{x}_v|^2)$$

$$= \frac{1}{4}(E + G)$$

$$= \frac{E}{2}.$$

Finally, by Exercise 2.4.17, $\frac{\partial \phi}{\partial \bar{z}} = \frac{\partial}{\partial \bar{z}}(\frac{\partial \mathbf{x}}{\partial z}) = \frac{1}{4}\Delta \mathbf{x} = 0$ since $\mathbf{x}$ is isothermal (and minimal). Therefore, each $\phi^i = \frac{\partial \mathbf{x}}{\partial z}$ is holomorphic. Conversely, the same calculation shows that, if each ϕ^i is holomorphic, then each x^i is harmonic and, therefore, M is minimal. All of these observations together give

Theorem 3.6.2. *Suppose M is a surface with parametrization $\mathbf{x}$. Let $\phi = \frac{\partial \mathbf{x}}{\partial z}$ and suppose $(\phi)^2 = 0$ (i.e. $\mathbf{x}$ is isothermal). Then M is minimal if and only if each ϕ^i is holomorphic.*

In case each ϕ^i is holomorphic we say that ϕ is holomorphic. The result above says that any minimal surface may be described near each of its points by a triple of holomorphic functions $\phi = (\phi^1, \phi^2, \phi^3)$ with $(\phi)^2 = 0$. Indeed, in this case we may construct an isothermal parametrization for a minimal surface by taking

Corollary 3.6.3. $x^i(z, \bar{z}) = c_i + 2 \operatorname{Re} \int \phi^i \, dz$.

Proof. Because $z = u + iv$, we may write $dz = du + idv$. (This is 'differential' shorthand for $\frac{dz}{dt} = \frac{du}{dt} + i\frac{dv}{dt}$.) Then

$$\phi^i \, dz = \frac{1}{2}[(x_u^i - ix_v^i)(du + idv)] = \frac{1}{2}[x_u^i \, du + x_v^i \, dv + i(x_u^i \, dv - x_v^i \, du)],$$

$$\bar{\phi}^i \, d\bar{z} = \frac{1}{2}[(x_u^i + ix_v^i)(du - idv)] = \frac{1}{2}[x_u^i \, du + x_v^i \, dv - i(x_u^i \, dv - x_v^i \, du)].$$

We then have

$$\begin{aligned} dx^i &= \frac{\partial x^i}{\partial z} dz + \frac{\partial x^i}{\partial \bar{z}} d\bar{z} \\ &= \phi^i \, dz + \bar{\phi}^i \, d\bar{z} \\ &= \phi^i \, dz + \overline{\phi^i \, dz} \\ &= 2\operatorname{Re} \phi^i \, dz, \end{aligned}$$

and we can now integrate to get x^i. $\qquad\qquad\square$

So, in a real sense, the problem of constructing minimal surfaces reduces to finding $\phi = (\phi^1, \phi^2, \phi^3)$ with $(\phi)^2 = 0$. A nice way of constructing such a ϕ is to take a holomorphic function f and a *meromorphic* function g (with fg^2 holomorphic) and form

$$\phi^1 = \frac{1}{2}f(1 - g^2), \qquad \phi^2 = \frac{i}{2}f(1 + g^2), \qquad \phi^3 = fg.$$

Exercise 3.6.4. Show that this ϕ satisfies $(\phi)^2 = 0$.

Exercise 3.6.5. Show that $g = \frac{\phi^3}{\phi^1 - i\phi^2}$.

Therefore, we obtain

Theorem 3.6.6 (Weierstrass-Enneper Representation I). *If f is holomorphic on a domain D, g is meromorphic on D and fg^2 is holomorphic on D, then a minimal surface is defined by the parametrization* $\mathbf{x}(z, \bar{z}) = \big(x^1(z, \bar{z}), x^2(z, \bar{z}), x^3(z, \bar{z})\big)$, *where*

$$x^1(z, \bar{z}) = \operatorname{Re} \int f(1 - g^2)\, dz,$$

$$x^2(z, \bar{z}) = \operatorname{Re} \int i\, f(1 + g^2)\, dz,$$

$$x^3(z, \bar{z}) = \operatorname{Re} 2\!\int fg\, dz.$$

Because we can write the coordinates of a parametrization in terms of integrals, we have a real way of creating minimal surfaces. In fact, with today's computer algebra systems, we can see our creations take shape before our eyes (see § 5.9).

To make things simpler yet, let's write another form of the Weierstrass-Enneper Representation so that we only need choose one function instead of two. Suppose, in Theorem 3.6.6, g is holomorphic and has an inverse function g^{-1} (in a domain D) which is holomorphic as well. Then we can consider g as a new complex variable $\tau = g$ with $d\tau = g'dz$. Define $F(\tau) = f/g'$ and obtain $F(\tau)d\tau = f\, dz$. Therefore, if we replace g by τ and $f\, dz$ by $F(\tau)d\tau$, we get

Theorem 3.6.7 (Weierstrass-Enneper Representation II). *For any holomorphic function $F(\tau)$, a minimal surface is defined by the parametrization* $\mathbf{x}(z, \bar{z}) = \big(x^1(z, \bar{z}), x^2(z, \bar{z}), x^3(z, \bar{z})\big)$, *where*

$$x^1(z, \bar{z}) = \operatorname{Re} \int (1 - \tau^2)F(\tau)\, d\tau,$$

$$x^2(z, \bar{z}) = \operatorname{Re} \int i\,(1 + \tau^2)F(\tau)\, d\tau,$$

$$x^3(z, \bar{z}) = \operatorname{Re} 2\!\int \tau F(\tau)\, d\tau.$$

Note the corresponding

$$\phi = \left(\frac{1}{2}(1 - \tau^2)F(\tau),\ \frac{i}{2}(1 + \tau^2)F(\tau),\ \tau F(\tau)\right).$$

This representation tells us that *any* holomorphic function $F(\tau)$ defines a minimal surface. Of course, we can't expect every function to give complex integrals which may be evaluated to nice formulas. Nevertheless, we shall see that we may calculate much information about a minimal surface directly from its representions.

Example 3.6.8 (The Helicoid). Let $F(\tau) = \frac{i}{2\tau^2}$. After integration, make the substitution $\tau = e^z$ and use the formulas in Exercise 2.4.1 to get

$$
\begin{aligned}
x^1 &= \operatorname{Re} \int (1 - \tau^2) \frac{i}{2\tau^2}\, d\tau \\
&= \operatorname{Re} \int \frac{i}{2\tau^2} - \frac{i}{2}\, d\tau \\
&= -\operatorname{Re} i \left[\frac{1}{2\tau} + \frac{\tau}{2} \right] \\
&= -\operatorname{Re} i \frac{e^{-z} + e^z}{2} \\
&= -\operatorname{Re}(i \cosh z) \\
&= \sinh u \sin v,
\end{aligned}
\qquad
\begin{aligned}
x^2 &= \operatorname{Re} i \int (1 + \tau^2) \frac{i}{2\tau^2}\, d\tau \\
&= \operatorname{Re} \int i \left[\frac{i}{2\tau^2} + \frac{i}{2} \right] d\tau \\
&= -\operatorname{Re} i^2 \left[\frac{1}{2\tau} - \frac{\tau}{2} \right] \\
&= \operatorname{Re} \frac{e^{-z} - e^z}{2} \\
&= \operatorname{Re}(-\sinh z) \\
&= -\sinh u \cos v,
\end{aligned}
$$

$$
\begin{aligned}
x^3 &= \operatorname{Re} 2 \int \tau \frac{i}{2\tau^2}\, d\tau \\
&= \operatorname{Re} \int \frac{i}{\tau}\, d\tau \\
&= \operatorname{Re} i \ln \tau \\
&= \operatorname{Re} i z \\
&= -v.
\end{aligned}
$$

The resulting parametrization $(\sinh u \sin v, -\sinh u \cos v, -v)$ is a form of the helicoid. (This helicoid is isometric to the usual one [**Opr97**], essentially because u and $\sinh u$ display the same global qualitative behavior.) Thus, the representation associated to $F(\tau)$ is a helicoid.

Exercise 3.6.9. Let $F(\tau) = \frac{1}{2\tau^2}$. Then, using the substitution $\tau = e^z$, obtain the Weierstrass-Enneper II representation for the catenoid

$$\mathbf{x}(u, v) = (-\cosh u \cos v, -\cosh u \sin v, u).$$

Exercise 3.6.10. Show that the catenoid and helicoid, respectively, also have representations of the form

$$(f, g) = (-\frac{e^{-z}}{2}, -e^z) \quad \text{and} \quad (f, g) = (-i\,\frac{e^{-z}}{2}, -e^z).$$

Exercise 3.6.11. Show that the helicoid has another representation $(f, g) = (-\frac{i}{2}, \frac{1}{z})$. This representation seems simpler than the first two, but there is a small problem. Namely, the integrals of the representation are not *path independent*. They are then said to have *real periods*. Show that if we carry out the integration for x^3 about the entire unit circle, we obtain a real period equal to 2π. To fix this ambiguity, we take a smaller domain $\mathbb{C}/\{0 \cup \mathbb{R}^-\}$. We therefore obtain one loop of the helicoid from this representation. Other loops may be similarly obtained and fitted together to create the whole surface. Thus, the simplicity of the representation $(f, g) = (-\frac{i}{2}, \frac{1}{z})$ hides the subtlety underlying complex integration.

Exercise 3.6.12. Let $F(\tau) = i(\frac{1}{\tau} - \frac{1}{\tau^3})$. Show that the associated representation is Catalan's surface

$$\mathbf{x}(u, v) = (u - \sin u \cosh v,\; 1 - \cos u \cosh v,\; 4 \sin \frac{u}{2} \sinh \frac{v}{2}).$$

Hint: after integrating, replace τ by $e^{-iz/2}$ and use the expansion of $\sin z$.

Exercise 3.6.13. Let $F(\tau) = 1 - \frac{1}{\tau^4}$. Show that the associated representation is Henneberg's surface

$$\mathbf{x}(u, v) = (2 \sinh u \cos v - \frac{2}{3} \sinh 3u \cos 3v,$$
$$2 \sinh u \sin v + \frac{2}{3} \sinh 3u \sin 3v,$$
$$2 \cosh 2u \cos 2v).$$

Let $v = \pi/2$ and show that you obtain Neil's parabola $(z-2)^3 = 9x^2$. This will determine Henneberg's surface in Example 3.9.16. Hint: for the first part, after integrating, replace τ by e^z, use the expansions of $\sin z$ and $\cos z$ and replace y by $-y$.

Exercise 3.6.14. Let $F(\tau) = 1$ or, equivalently, $(f, g) = (1, z)$. Show that the Weierstrass-Enneper representations give you Enneper's surface. If $(f, g) = (1, z^n)$, then an nth-order Enneper surface is obtained. Calculate the $\mathbf{x}(u, v)$ for the second-order Enneper surface. Hint: Maple might be useful here. See § 5.9.

Exercise 3.6.15. Let $(f, g) = (z^2, \frac{1}{z^2})$. Calculate the $\mathbf{x}(u, v)$ given by the Weierstrass-Enneper representation. This surface is called *Richmond's surface*.

Exercise 3.6.16. Let $F(\tau) = \frac{2}{1-\tau^4}$. Show that the Weierstrass-Enneper representation gives Scherk's surface $z = \ln(\frac{\cos y}{\cos x})$. Hints: (1) for x^1, use partial fractions to get $\int \frac{2}{1+\tau^2} d\tau = i\log(1 - i\tau) - i\log(1 + i\tau)$, (2) since the log's are multiplied by i and you want the real part, the definition of $\log(z)$ gives $x^1 = \arctan(\frac{u}{1-v}) - \arctan(\frac{-u}{1+v})$ (i.e. $1 + i\tau = 1 - v + iu$ and $1 - i\tau = 1 + v - iu$), (3) the formula for the tangent of a difference of two angles then gives $x^1 = \arctan(\frac{2u}{1-(u^2+v^2)})$, (4) similarly, $x^2 = \arctan(\frac{-2v}{1-(u^2+v^2)})$, (5) also, $x^3 = \text{Re}(\log(\tau^2 + 1) - \log(\tau^2 - 1)) = \frac{1}{2}\ln\left(\frac{(u^2-v^2+1)^2+4u^2v^2}{(u^2-v^2-1)^2+4u^2v^2}\right)$, (6) by drawing appropriate right triangles, write $\cos x^2$ and $\cos x^3$ in terms of u and v and show $x^3 = \ln(\frac{\cos x^2}{\cos x^1})$. Note that the integrals have real periods so that, just as for the helicoid, we can take one piece at a time to build up the whole surface.

The real value of the Weierstrass-Enneper representations is that we can analyze many facets of minimal surfaces directly from the representing functions (f, g) and $F(\tau)$. This applies even to surfaces whose Weierstrass-Enneper integrals may not be explicitly computable. As an example of this, we shall compute the Gauss curvature

K of a minimal surface in terms of $F(\tau)$. First note that, by Theorem 2.5.3, isothermal parameters give

$$
\begin{aligned}
K &= -\frac{1}{2\sqrt{EG}} \left(\frac{\partial}{\partial v} \left(\frac{E_v}{\sqrt{EG}} \right) + \frac{\partial}{\partial u} \left(\frac{G_u}{\sqrt{EG}} \right) \right) \\
&= -\frac{1}{2E} \left(\frac{\partial}{\partial v} \left(\frac{E_v}{E} \right) + \frac{\partial}{\partial u} \left(\frac{E_u}{E} \right) \right) \\
&= -\frac{1}{2E} \left(\frac{\partial^2}{\partial v^2} \ln E + \frac{\partial^2}{\partial u^2} \ln E \right) \\
&= -\frac{1}{2E} \Delta(\ln E),
\end{aligned}
$$

where Δ is the Laplace operator $\frac{\partial^2}{\partial u^2} + \frac{\partial^2}{\partial v^2}$. By Exercise 3.6.1, we know that $E = 2|\phi|^2$, so we compute E for

$$
\phi = (\frac{1}{2}(1 - \tau^2)F(\tau), \frac{i}{2}(1 + \tau^2)F(\tau), \tau F(\tau)).
$$

We have

$$
\begin{aligned}
E &= 2\left[\left| \frac{1}{2}(1 - \tau^2)F(\tau) \right|^2 + \left| \frac{i}{2}(1 + \tau^2)F(\tau) \right|^2 + |\tau F(\tau)|^2 \right] \\
&= \frac{1}{2}|F|^2 \left[|\tau^2 - 1|^2 + |\tau^2 + 1|^2 + 4|\tau|^2 \right].
\end{aligned}
$$

Now, $\tau^2 = u^2 - v^2 + 2iuv$, so $|\tau^2 - 1|^2 = (u^2 - v^2 - 1)^2 + 4u^2v^2$. Similarly, $|\tau^2 + 1|^2 = (u^2 - v^2 + 1)^2 + 4u^2v^2$ and $4|\tau|^2 = 4(u^2 + v^2)$. Then

$$
\begin{aligned}
E &= \frac{1}{2}|F|^2 2 \left[(u^2 - v^2)^2 + 1 + 4u^2v^2 + 2u^2 + 2v^2 \right]. \\
&= |F|^2 \left[u^4 + 2u^2v^2 + v^4 + 1 + 2u^2 + 2v^2 \right] \\
&= |F|^2 \left[1 + u^2 + v^2 \right]^2.
\end{aligned}
$$

Exercise 3.6.17. Show directly from the definition of ϕ in terms of the representation (f, g) that $E = |f|^2 \left[1 + |g|^2 \right]^2$. Hint: use Exercise 3.6.1 and the equality

$$
|\phi|^2 = \frac{1}{4}|f|^2((1 - g^2)(1 - \bar{g}^2) + (1 + g^2)(1 + \bar{g}^2) + 4g\bar{g}).
$$

Now, $\ln E = \ln |F|^2 + 2\ln(1 + u^2 + v^2)$, and it is an easy task to prove the following.

Exercise 3.6.18. Show that
$$\Delta(2\ln(1 + u^2 + v^2)) = \frac{8}{(1 + u^2 + v^2)^2}.$$

We must now calculate
$$\Delta(\ln |F|^2) = \Delta(\log F\bar{F}) = \Delta(\log F + \log \bar{F}).$$

Previously, we have seen that $\Delta = 4\partial^2/\partial\bar{z}\partial z$. Further, because F is holomorphic, $\bar{F}$ cannot be, so $\partial\bar{F}/\partial z = 0$ and, consequently, $\partial(\log \bar{F})/\partial z = 0$. We are left with
$$\Delta(\log F) = 4\frac{\partial^2(\log F)}{\partial\bar{z}\partial z} = 4\frac{\partial(F'/F)}{\partial\bar{z}} = 0,$$

since F, F' and, hence, F'/F are holomorphic. Thus, $\Delta(\ln |F|^2) = 0$ and $\Delta(\ln E) = 8/(1 + u^2 + v^2)^2$.

Theorem 3.6.19. *The Gauss curvature of the minimal surface determined by the Weierstrass-Enneper representation II is*
$$K = \frac{-4}{|F|^2(1 + u^2 + v^2)^4}.$$

Proof. From the calculations above,
$$K = -\frac{1}{2E}\Delta(\ln E)$$
$$= \frac{-8}{2|F|^2(1 + u^2 + v^2)^4}$$
$$= \frac{-4}{|F|^2(1 + u^2 + v^2)^4}.$$

$\square$

Exercise 3.6.20. Use the identification $F = \frac{f}{g'}$ to derive the formula
$$K = \frac{-4|g'|^2}{|f|^2(1 + |g|^2)^4}$$
in terms of the Weierstrass-Enneper representation I.

Exercise 3.6.21. Explain the apparent discrepancy between the two formulas for K. That is, the first formula never allows $K = 0$ while the second has $K = 0$ at points where $g' = 0$. Hint: what was our assumption about g which allowed the transformation from Weierstrass-Enneper I to Weierstrass-Enneper II?

Exercise 3.6.22. A minimal surface described by (f, g) or $F(\tau)$ has an *associated family* of minimal surfaces given by (respectively) $(e^{it}f, g)$ or $e^{it}F(\tau)$. Two surfaces of the family described by t_0 and t_1 are said to be *adjoint* if $t_1 - t_0 = \pi/2$. Show that E remains the same no matter what t is taken. Because we have isothermal parameters, this is enough to show that all the surfaces of an associated family are locally isometric. In fact, Schwarz proved that if two minimal surfaces are isometric, then one of them must be congruent to some member of an associated family of the other (see [**Spi79**, vol. 4, Ch. 9]). Also see Remark 3.2.9 and § 5.7.

Exercise 3.6.23. The catenoid has a Weierstrass-Enneper representation $(f, g) = (-\frac{e^{-z}}{2}, -e^z)$. Write the $\mathbf{x}(u, v)$ for its associated family and its adjoint surface.

Exercise 3.6.24. Find the adjoint surface to Henneberg's surface with $F(\tau) = -i(1 - \frac{1}{\tau^4})$. Set $u = 0$ and show that the resulting v-parameter curve is the astroid $x^{2/3} + y^{2/3} = (8/3)^{2/3}$. This astroid is a geodesic. After doing Exercise 3.9.16, can you do the same for the astroid in this surface?

The Weierstrass-Enneper representations tell us that minimal surface theory sits at the feet of complex analysis. Nevertheless, it is always helpful in mathematics to look at the same object from many viewpoints, so let's return to some fundamental differential geometry.

3.7. The Gauss Map

The *Gauss map* of a surface M with parametrization $\mathbf{x}(u, v)$ is a mapping from the surface M to the unit sphere $S^2 \subset \mathbb{R}^3$, denoted

$G: M \to S^2$ and given by $G(p) = U_p$, where U_p is the unit normal to M at p. In terms of the parametrization, we may write $G(\mathbf{x}(u,v)) = U(u,v)$ and, for a small piece of M, think of $U(u,v)$ as a parametrization of part of the sphere S^2. There is also an induced linear transformation of tangent planes given, for the basis $\{\mathbf{x}_u, \mathbf{x}_v\}$, by $G_*(\mathbf{x}_u) = U_u$ and $G_*(\mathbf{x}_v) = U_v$. To understand this, note the following. A tangent vector on M is the velocity vector of some curve on M, so by taking U only along the curve, we create a new curve on the sphere S^2. The tangent vector of the new spherical curve is then, by definition, the image under G_* of the original curve's tangent vector. Applying this reasoning to the parameter curves, we see that $G_*(\mathbf{x}_u) = U_u$ and $G_*(\mathbf{x}_v) = U_v$.

The Gauss map of any surface is intimately tied up with the Gauss curvature. For a discussion of this, see [**dC76**] for instance.

Exercise 3.7.1. Calculate the following quantity using the definition of the Gauss map and Formula(s) 2.5.10:

$$\frac{|G_*(\mathbf{x}_u) \times G_*(\mathbf{x}_v)|}{|\mathbf{x}_u \times \mathbf{x}_v|}.$$

What is this saying geometrically? Think of small pieces of area.

But there is something even more special about the Gauss map of a minimal surface. In order to state the result, we need to recall that a linear transformation $T: \mathbb{R}^2 \to \mathbb{R}^2$ is *conformal* if, for fixed $\rho > 0$,

$$T(x) \cdot T(y) = \rho^2\, x \cdot y$$

for all vectors $x, y \in \mathbb{R}^2$. (There is a more general notion, but this is all we will need.) We also have the following

Proposition 3.7.2. *T is conformal if and only if, for a basis $\{v_1, v_2\}$ of $\mathbb{R}^2$, $|T(v_1)| = \rho|v_1|$, $|T(v_2)| = \rho|v_2|$ and $T(v_1) \cdot T(v_2) = \rho^2\, v_1 \cdot v_2$ for $\rho > 0$.*

Proof. The implication ($\Rightarrow$) is trivial. For ($\Leftarrow$), write $x = av_1 + bv_2$ and $y = cv_1 + dv_2$ with $x \cdot y = ac|v_1|^2 + (bc + ad)v_1 \cdot v_2 + bd|v_2|^2$. The linearity of T gives $T(x) = aT(v_1) + bT(v_2)$ and $T(y) = cT(v_1) +$

$dT(v_2)$ with

$$
\begin{aligned}
T(x) \cdot T(y) &= ac|T(v_1)|^2 + (bc + ad)T(v_1) \cdot T(v_2) + bd|T(v_2)|^2 \\
&= ac\rho^2|v_1|^2 + (bc + ad)\rho^2 v_1 \cdot v_2 + bd\rho^2|v_2|^2 \\
&= \rho^2\, x \cdot y.
\end{aligned}
$$

$\square$

Exercise 3.7.3. Show that if T is conformal, then angles between tangent vectors are preserved. (Hint: recall that $x \cdot y = |x||y|\,\cos(\theta)$.)

When $F\colon M \to N$ is a mapping of one surface to another, F is said to be *conformal* when the induced map on tangent vectors F_* is conformal for each point of M. In this case, the factor ρ varies from point to point and is, therefore, a function of the surface parameters u and v. We then write $\rho(u, v)$ and call it the *scaling factor*. On each tangent plane, however, ρ is constant.

Example 3.7.4 (Gauss Map of the Helicoid). Consider the helicoid $\mathbf{x}(u, v) = (u \cos v, u \sin v, v)$. We must compute $G_*(\mathbf{x}_u) = U_u$ and $G_*(\mathbf{x}_v) = U_v$. We have

$$
\mathbf{x}_u = (\cos v, \sin v, 0), \qquad \mathbf{x}_v = (-u \sin v, u \cos v, 1)
$$

with $\mathbf{x}_u \cdot \mathbf{x}_v = 0$ and $|\mathbf{x}_u| = 1$ and $|\mathbf{x}_v| = \sqrt{1 + u^2}$. The unit normal is given by

$$
U = \frac{(\sin v, -\cos v, u)}{\sqrt{1 + u^2}}.
$$

Taking partial derivatives gives

$$
U_u = \frac{(-u \sin v, u \cos v, 1)}{(1 + u^2)^{3/2}}, \qquad U_v = \frac{(\cos v, \sin v, 0)}{\sqrt{1 + u^2}}
$$

with lengths computed to be

$$
|U_u| = \frac{1}{1 + u^2} = \frac{1}{1 + u^2}|\mathbf{x}_u|,
$$

$$
|U_v| = \frac{1}{\sqrt{1 + u^2}} = \frac{1}{1 + u^2}|\mathbf{x}_v|.
$$

Also, of course, $G_*(\mathbf{x}_u) \cdot G_*(\mathbf{x}_v) = U_u \cdot U_v = 0$, as required for the Gauss map G to be conformal with scaling factor $\rho(u, v) = 1/(1+u^2)$.

Exercise 3.7.5. For the catenoid and Enneper's surface,

$$\mathbf{x}(u,v) = (\cosh u \cos v, \cosh u \sin v, u),$$
$$\mathbf{x}(u,v) = (u - u^3/3 + uv^2, -v + v^3/3 - vu^2, u^2 - v^2),$$

respectively, show that the Gauss maps are conformal with respective scaling factors $1/\cosh^2 u$, $2/(1+u^2+v^2)^2$. For Enneper, also see § 5.8.

Example 3.7.4 and Exercise 3.7.5 are not just isolated instances, as the following result shows.

Proposition 3.7.6. *Let M be a minimal surface with isothermal parametrization $\mathbf{x}(u,v)$. Then the Gauss map of M is a conformal map.*

Proof. In order to show G to be conformal, we need only show $|G_*(\mathbf{x}_u)| = \rho(u,v)|\mathbf{x}_u|$, $|G_*(\mathbf{x}_v)| = \rho(u,v)|\mathbf{x}_v|$ and $G_*(\mathbf{x}_u) \cdot G_*(\mathbf{x}_v) = \rho^2(u,v)\,\mathbf{x}_u \cdot \mathbf{x}_v$. Since isothermal parameters have $E = G$ and $F = 0$, we get $H = (\ell + n)/2E$ as well as

$$G_*(\mathbf{x}_u) = U_u = -\frac{\ell}{E}\mathbf{x}_u - \frac{m}{E}\mathbf{x}_v,$$
$$G_*(\mathbf{x}_v) = U_v = -\frac{m}{E}\mathbf{x}_u - \frac{n}{E}\mathbf{x}_v,$$

where we have used the formulas for U_u and U_v in Formula(s) 2.5.10. Using isothermal coordinates and taking dot products gives

$$|U_u|^2 = \frac{1}{E}\left[\ell^2 + m^2\right], \qquad |U_v|^2 = \frac{1}{E}\left[m^2 + n^2\right],$$

$$U_u \cdot U_v = \frac{m}{E}\left[\ell + n\right].$$

But M is minimal, so $H = 0$ and this means $\ell = -n$ by Exercise 3.4.3. We then obtain

$$|U_u|^2 = \frac{1}{E}\left[\ell^2 + m^2\right] = |U_v|^2 \quad \text{and} \quad U_u \cdot U_v = 0.$$

Since $|\mathbf{x}_u| = \sqrt{E} = |\mathbf{x}_v|$ and $\mathbf{x}_u \cdot \mathbf{x}_v = 0$, we see that the Gauss map G is conformal with scaling factor $\sqrt{\ell^2 + m^2}/E$. $\qquad\square$

Exercise 3.7.7. Show that the scaling factor $\sqrt{\ell^2 + m^2}/E$ is equal to $\sqrt{|K|}$, where K is the Gauss curvature. Hint: $K = (-\ell^2 - m^2)/E^2$. Why?

The following result shows that conformality of the Gauss map almost characterizes minimal surfaces.

Proposition 3.7.8. *Let M be a surface parametrized by $\mathbf{x}(u, v)$ whose Gauss map $G\colon M \to S^2$ is conformal. Then either M is (part of) a sphere or M is a minimal surface.*

Proof. Because we've only developed formulas for surfaces with $F = 0$, assume that the parametrization $\mathbf{x}(u, v)$ has this property. Because the Gauss map is conformal and $F = \mathbf{x}_u \cdot \mathbf{x}_v = 0$, then we get $U_u \cdot U_v = 0$ also. By Formula(s) 2.5.10, this implies that $m(G\ell + En) = 0$. Therefore, either $m = 0$ or $G\ell + En = 0$. Since $F = 0$, the second condition really says that $H = 0$ and M is minimal. If this condition doesn't hold, then m must be zero. Now use $m = 0$, conformality and Formula(s) 2.5.10 again to get

$$\frac{\ell^2}{E} = U_u \cdot U_u = \rho^2 E, \qquad \frac{n^2}{G} = U_v \cdot U_v = \rho^2 G.$$

Multiplying across each equation produces

$$\frac{\ell^2}{E^2} = \rho^2 = \frac{n^2}{G^2} \;\Rightarrow\; \frac{\ell}{E} = \pm\frac{n}{G}.$$

Suppose $\ell/E = -n/G$. Then finding a common denominator shows again that $H = 0$ and M is minimal. Suppose that $\ell/E = n/G$ at every point of M, and denote this common value by k. Together with $m = 0$, this says that $-U_u = k\,\mathbf{x}_u$ and $-U_v = k\,\mathbf{x}_v$. These equalities show that, at each point of M, the tangent vectors $\mathbf{x}_u$ and $\mathbf{x}_v$ are eigenvectors of the operator (called the shape operator) defined by taking a kind of directional derivative (called the covariant derivative) of the unit normal. The eigenvalues associated with the eigenvectors of the shape operator are known to be the maximum and minimum normal curvatures at a point. For the situation above, both eigenvalues are equal to k, so all normal curvatures at a point are constant. Such a point on a surface is called an umbilic point. Therefore, when we suppose $m = 0$ and $\ell/E = n/G$, we are saying that every point of

M is an umbilic point. Such a non-planar surface is known to be a sphere. For all the differential geometry mentioned here, see, for example, [**Opr97**, Chapter 3] (especially [**Opr97**, Theorem 3.5.1]). □

Exercise 3.7.9. Verify all the statements in the proof of Proposition 3.7.8. Show that any umbilic point of a minimal surface must have zero Gauss curvature there. Such points are called *flat*. Consequently, for a minimal surface described by the Weierstrass-Enneper representation I with data (f, g), umbilic points are exactly those points where g' vanishes. Finally, this says that umbilic points are isolated on the surface. Hints: (1) umbilic means that all normal curvatures at a point are k; (2) $H = 0$, so what does this say about normal curvature (see Definition 2.3.1); (3) Gauss curvature is the product of the maximum and minimum normal curvatures; (4) Theorem 3.6.20; (5) Theorem 2.4.18.

3.8. Stereographic Projection and the Gauss Map

The Gauss map for a minimal surface has a description directly in terms of the Weierstrass-Enneper representation. This will prove important later (see Theorem 3.10.6) when we try to understand whether minimal surfaces always minimize surface area. *Stereographic projection from the North pole N* is denoted by

$$\text{St}: S^2/\{N\} \to \mathbb{R}^2$$

and defined by

$$\text{St}(\cos u \cos v, \sin u \cos v, \sin v) = \left(\frac{\cos u \cos v}{1 - \sin v}, \frac{\sin u \cos v}{1 - \sin v}, 0 \right).$$

As we have been doing, we can take the induced mapping on tangent vectors by differentiating with respect to u and v:

$$\text{St}_*(\mathbf{x}_u) = \left(\frac{-\sin u \cos v}{1 - \sin v}, \frac{\cos u \cos v}{1 - \sin v}, 0 \right),$$

$$\text{St}_*(\mathbf{x}_v) = \left(\frac{\cos u}{1 - \sin v}, \frac{\sin u}{1 - \sin v}, 0 \right).$$

Hence, taking dot products in $\mathbb{R}^3$, we get

$$\mathrm{St}_*(\mathbf{x}_u) \cdot \mathrm{St}_*(\mathbf{x}_u) = \frac{\cos^2 v}{(1 - \sin v)^2}$$

$$= \frac{1}{(1 - \sin v)^2} \mathbf{x}_u \cdot \mathbf{x}_u,$$

$$\mathrm{St}_*(\mathbf{x}_v) \cdot \mathrm{St}_*(\mathbf{x}_v) = \frac{1}{(1 - \sin v)^2}$$

$$= \frac{1}{(1 - \sin v)^2} \mathbf{x}_v \cdot \mathbf{x}_v,$$

$$\mathrm{St}_*(\mathbf{x}_u) \cdot \mathrm{St}_*(\mathbf{x}_v) = 0.$$

The factor $1/(1 - \sin v)$ shows that stereographic projection is a conformal map with scaling factor $1/(1 - \sin v)$. That is, in particular, stereographic projection preserves angles (see Exercise 3.7.3). In Cartesian coordinates, stereographic projection is given by

$$\mathrm{St}(x, y, z) = (x/(1 - z), y/(1 - z), 0).$$

We may identify the real plane $\mathbb{R}^2$ with the complex plane $\mathbb{C}$ and extend St to a one-to-one onto mapping $\mathrm{St} \colon S^2 \to \mathbb{C} \cup \{\infty\}$ with the North pole mapping to ∞. With these identifications, we have

Theorem 3.8.1. *Let M be a minimal surface with isothermal parametrization $\mathbf{x}(u, v)$ and Weierstrass-Enneper representation (f, g). Then the Gauss map of M, $G \colon M \to \mathbb{C} \cup \{\infty\}$, may be identified with the meromorphic function g.*

Proof. Recall that $\phi = \frac{\partial \mathbf{x}}{\partial z}$, $\bar{\phi} = \frac{\partial \mathbf{x}}{\partial \bar{z}}$ and

$$\phi^1 = \frac{1}{2} f(1 - g^2), \quad \phi^2 = \frac{i}{2} f(1 + g^2), \quad \phi^3 = fg.$$

We will describe the Gauss map in terms of ϕ^1, ϕ^2 and ϕ^3. First, we write

$$\mathbf{x}_u \times \mathbf{x}_v = ((\mathbf{x}_u \times \mathbf{x}_v)^1, \ (\mathbf{x}_u \times \mathbf{x}_v)^2, \ (\mathbf{x}_u \times \mathbf{x}_v)^3)$$

$$= (x_u^2 x_v^3 - x_u^3 x_v^2, \ x_u^3 x_v^1 - x_u^1 x_v^3, \ x_u^1 x_v^2 - x_u^2 x_v^1),$$

and consider the first component $(\mathbf{x}_u \times \mathbf{x}_v)^1 = x_u^2 x_v^3 - x_u^3 x_v^2$. We have

$$
\begin{aligned}
x_u^2 x_v^3 - x_u^3 x_v^2 &= \mathrm{Im}[(x_u^2 - i x_v^2)(x_u^3 + i x_v^3)] \\
&= \mathrm{Im}[2(\partial x^2/\partial z) \cdot 2(\partial x^3/\partial \bar{z})] \\
&= 4\,\mathrm{Im}(\phi^2 \bar{\phi}^3).
\end{aligned}
$$

Similarly, $(\mathbf{x}_u \times \mathbf{x}_v)^2 = 4\,\mathrm{Im}(\phi^3 \bar{\phi}^1)$ and $(\mathbf{x}_u \times \mathbf{x}_v)^3 = 4\,\mathrm{Im}(\phi^1 \bar{\phi}^2)$. Hence, we obtain

$$
\mathbf{x}_u \times \mathbf{x}_v = 4\,\mathrm{Im}(\phi^2 \bar{\phi}^3, \phi^3 \bar{\phi}^1, \phi^1 \bar{\phi}^2) = 2(\phi \times \bar{\phi}),
$$

where the last equality follows from $z - \bar{z} = 2\,\mathrm{Im}\,z$. Now, since $\mathbf{x}(u, v)$ is isothermal, $|\mathbf{x}_u \times \mathbf{x}_v| = |\mathbf{x}_u| \cdot |\mathbf{x}_v| = |\mathbf{x}_u|^2 = E = 2|\phi|^2$ by Exercise 3.6.1. Therefore, we have

$$
U = \frac{\mathbf{x}_u \times \mathbf{x}_v}{|\mathbf{x}_u \times \mathbf{x}_v|} = \frac{2(\phi \times \bar{\phi})}{2|\phi|^2} = \frac{\phi \times \bar{\phi}}{|\phi|^2}.
$$

The Gauss map $G \colon M \to \mathbb{C} \cup \{\infty\}$ may now be given in terms of the ϕ^i:

$$
\begin{aligned}
G(\mathbf{x}(u, v)) &= \mathrm{St}(U(u, v)) \\
&= \mathrm{St}\left(\frac{\phi \times \bar{\phi}}{|\phi|^2}\right) \\
&= \mathrm{St}\left(\frac{2\,\mathrm{Im}(\phi^2 \bar{\phi}^3, \phi^3 \bar{\phi}^1, \phi^1 \bar{\phi}^2)}{|\phi|^2}\right) \\
&= \left(\frac{2\,\mathrm{Im}(\phi^2 \bar{\phi}^3)}{|\phi|^2 - 2\,\mathrm{Im}(\phi^1 \bar{\phi}^2)}, \frac{2\,\mathrm{Im}(\phi^3 \bar{\phi}^1)}{|\phi|^2 - 2\,\mathrm{Im}(\phi^1 \bar{\phi}^2)}, 0\right).
\end{aligned}
$$

The last equality follows because

$$
\begin{aligned}
\frac{x}{1 - z} &= \frac{2\,\mathrm{Im}(\phi^2 \bar{\phi}^3)}{|\phi|^2} \cdot \frac{1}{1 - \frac{2\,\mathrm{Im}(\phi^1 \bar{\phi}^2)}{|\phi|^2}} \\
&= \frac{2\,\mathrm{Im}(\phi^2 \bar{\phi}^3)}{|\phi|^2} \cdot \frac{|\phi|^2}{|\phi|^2 - 2\,\mathrm{Im}(\phi^1 \bar{\phi}^2)} \\
&= \frac{2\,\mathrm{Im}(\phi^2 \bar{\phi}^3)}{|\phi|^2 - 2\,\mathrm{Im}(\phi^1 \bar{\phi}^2)}
\end{aligned}
$$

and similarly for $y/(1-z)$. Identifying $(x, y) \in \mathbb{R}^2$ with $x + iy \in \mathbb{C}$ allows us to write

$$G(\mathbf{x}(u,v)) = \frac{2\,\mathrm{Im}(\phi^2\bar{\phi}^3) + 2i\,\mathrm{Im}(\phi^3\bar{\phi}^1)}{|\phi|^2 - 2\,\mathrm{Im}(\phi^1\bar{\phi}^2)}.$$

Now, let's consider the numerator $\mathcal{N}$ of this fraction:

$$\begin{aligned}
\mathcal{N} &= 2\,\mathrm{Im}(\phi^2\bar{\phi}^3) + 2i\,\mathrm{Im}(\phi^3\bar{\phi}^1) \\
&= \frac{1}{i}[\phi^2\bar{\phi}^3 - \bar{\phi}^2\phi^3 + i\,\phi^3\bar{\phi}^1 - i\,\bar{\phi}^3\phi^1] \\
&= \phi^3(\bar{\phi}^1 + i\,\bar{\phi}^2) - \bar{\phi}^3(\phi^1 + i\,\phi^2).
\end{aligned}$$

Also, $0 = (\phi)^2 = (\phi^1)^2 + (\phi^2)^2 + (\phi^3)^2 = (\phi^1 - i\,\phi^2)(\phi^1 + i\,\phi^2) + (\phi^3)^2$, so

$$\phi^1 + i\,\phi^2 = \frac{-(\phi^3)^2}{\phi^1 - i\,\phi^2}.$$

Then we have

$$\begin{aligned}
\mathcal{N} &= \phi^3(\bar{\phi}^1 + i\,\bar{\phi}^2) + \bar{\phi}^3\,\frac{(\phi^3)^2}{\phi^1 - i\,\phi^2} \\
&= \frac{\phi^3[(\phi^1 - i\,\phi^2)(\bar{\phi}^1 + i\,\bar{\phi}^2) + |\phi^3|^2]}{\phi^1 - i\,\phi^2} \\
&= \frac{\phi^3}{\phi^1 - i\,\phi^2}\left[|\phi^1|^2 + |\phi^2|^2 + |\phi^3|^2 + i(\bar{\phi}^2\phi^1 - \phi^2\bar{\phi}^1)\right] \\
&= \frac{\phi^3}{\phi^1 - i\,\phi^2}\left[|\phi|^2 - 2\,\mathrm{Im}(\phi^1\bar{\phi}^2)\right].
\end{aligned}$$

The second factor of the numerator $\mathcal{N}$ cancels the denominator of $G(\mathbf{x}(u,v))$, and we end up with

$$G(\mathbf{x}(u,v)) = \frac{\phi^3}{\phi^1 - i\,\phi^2}.$$

By Exercise 3.6.5, we know that $g = \frac{\phi^3}{\phi^1 - i\,\phi^2}$ as well, so we are done. $\square$

Remark 3.8.2. Using the Weierstrass-Enneper representation II, we see that the Gauss map may be identified with the complex variable τ as well.

Exercise 3.8.3. Work through the following outline of *Bernstein's Theorem*: If a minimal surface $M: z = f(x,y)$ is defined on the whole xy-plane, then M is a plane. Hints: (1) In the proof of the existence of isothermal coordinates, if the parameter domain is the whole plane, then the potential functions μ and ρ can be extended over the plane as well. The mapping T then becomes a diffeomorphism (i.e. a smooth one-to-one onto map with smooth inverse) between the xy- and uv-planes. Therefore, we may assume that M has parameter domain the whole uv-plane, where u and v are isothermal coordinates. (2) The normals for M are contained in a hemisphere (by the Gauss map). Rotate the sphere to get them in the lower hemisphere. (3) Think of the uv-plane as the complex plane $\mathbb{C}$ and look at the composition $\mathbb{C} \to M \xrightarrow{G} S^2/\{N\} \to \mathbb{C}$. Why is this map holomorphic? Think of g. (4) Liouville's Theorem in complex analysis says that a complex function defined on the entire complex plane which is both bounded and holomorphic is constant.

Stereographic projection can also be used together with Exercise 3.6.20 to study the total Gauss curvature $\int_M K$ of a minimal surface (see [**Oss86**, §9]). In particular, the following result can be shown using these ideas.

Theorem 3.8.4. *The catenoid and Enneper's surface are the only minimal surfaces whose Gauss map is one-to-one.*

3.9. Creating Minimal Surfaces from Curves

The Weierstrass-Enneper representations *create* minimal surfaces from holomorphic functions. While this is analytically beautiful, a geometer might hope that a more geometric condition could be used to create minimal surfaces. For example, in Plateau's problem, we desire a minimal surface which spans a given boundary curve. In this case, preassigning the values of the minimal surface on the boundary puts a considerable constraint on construction of the surface. There are other natural constraints, however, which are easier to handle, but no less interesting. For example, suppose we wish to construct a minimal surface which contains a given curve as a parameter curve.

With a little extra information, this can be done; and this section is devoted to showing exactly how it can be done. Recall that a *vector field* $\mathcal{N}$ along a curve $\alpha\colon I \to \mathbb{R}^3$ is a smooth map $\mathcal{N}\colon I \to \mathbb{R}^3$ defined on the same domain as α. We think of the vectors $\mathcal{N}(t)$ as originating from the points in $\mathbb{R}^3$ belonging to the image of α. We will mention specific vector fields below, but we point out here that we need the vector fields (and curves) to be real-analytic. That is, the coordinate functions must have Taylor series which converge to the function's value at each point of the domain. In practice, everything we deal with has this property, so the reader should not fret unduly. So, let $\alpha(t)$ be a curve in $\mathbb{R}^3$ and let $\mathcal{N}(t)$ be a vector field along α which has the extra property that $\mathcal{N}(t) \cdot \alpha'(t) = 0$ for all $t \in I$, where $\alpha'(t)$ is the tangent vector to α at t.

Problem 3.9.1 (Björling's Problem). *Given a real analytic curve $\alpha\colon I \to \mathbb{R}^3$ and a real analytic vector field $\mathcal{N}$ along α with $\mathcal{N} \cdot \alpha' = 0$, construct a minimal surface M with parametrization $\mathbf{x}(u, v)$ having u-parameter curve $\alpha(u) = \mathbf{x}(u, 0)$ such that the vector field $\mathcal{N}$ along α is the unit normal of M, $\mathcal{N}(u) = U(u, 0)$, for $u \in I$.*

To solve Björling's problem, we make repeated use of the Identity theorem Theorem 2.4.18. Also, here we can note that the curve α and vector field $\mathcal{N}$ are required to be real analytic because we can convert such curves and vector fields automatically to their complex counterparts by replacing the real variable in the Taylor series of α and $\mathcal{N}$ by a complex variable z. The resulting complex Taylor series defines a holomorphic complex curve $\alpha(z)\colon D \to \mathbb{C}^3$ on some domain $D \subseteq \mathbb{C}$ containing the interval I. We say that $\alpha(z)$ is a *holomorphic extension* of $\alpha(t)$.

Given $\alpha(t)$ and $\mathcal{N}(t)$ as above, let $\alpha(z)$ and $\mathcal{N}(z)$ be the respective holomorphic extensions on a simply connected domain D containing I. We then have

Theorem 3.9.2. *There is exactly one solution to Björling's problem and it is given by*

$$\mathbf{x}(u, v) = \mathrm{Re}\left[\alpha(z) - i\int_{u_0}^{z} \mathcal{N}(w) \times \alpha'(w)\, dw\right],$$

where $u_0 \in I$ is fixed and $z = u + iv$.

Proof. We start by proving that the solution is unique if it exists. Suppose $\mathbf{x}(u,v)$ is a solution to Björling's problem. That is, $\mathbf{x}(u,v)$ is a minimal surface in isothermal parameters having $\mathbf{x}(u,0) = \alpha(u)$ and $U(u,0) = N(u)$. Since $\mathbf{x}(u,v)$ is isothermal, each coordinate function $x^j(u,v)$ is harmonic. Let $y^j(u,v)$ denote the corresponding harmonic conjugate (with $y^j(u_0,0) = 0$ say) and note that $x^j(z) + iy^j(z)$ is holomorphic. We can create a holomorphic curve $\beta\colon D \to \mathbb{C}^3$ by defining, for $z = u + iv \in D$,

$$\beta(z) = \mathbf{x}(u,v) + i\,\mathbf{y}(u,v).$$

Now, $\beta'(z) = \mathbf{x}_u + i\,\mathbf{y}_u = \mathbf{x}_u - i\,\mathbf{x}_v$ by § 2.4. But $\mathbf{x}(u,v)$ is in isothermal parameters, so $\mathbf{x}_u$, $\mathbf{x}_v$, $U = \mathbf{x}_u \times \mathbf{x}_v / |\mathbf{x}_u \times \mathbf{x}_v|$ are mutually orthogonal and $|\mathbf{x}_u| = |\mathbf{x}_v|$. Therefore, we have $\mathbf{x}_v = U \times \mathbf{x}_u$ and

$$\beta' = \mathbf{x}_u - i\,U \times \mathbf{x}_u.$$

If we restrict to $(u,0) \in I$, then we have $\beta'(u) = \alpha'(u) - i\,N(u) \times \alpha'(u)$ since $\mathbf{x}_u(u,0) = \alpha'(u)$ and $U(u,0) = N(u)$. We integrate this equation in the usual real variables sense to get

$$\beta(u) = \alpha(u) - i \int_{u_0}^{u} N(t) \times \alpha'(t)\,dt$$

for all $u \in I$. But now we see that, on the whole interval I, β agrees with the holomorphic curve $\gamma(z) = \alpha(z) - i \int N(w) \times \alpha'(w)\,dw$. The Identity theorem then says that β and γ agree on all of D. Since the real part of β is $\mathbf{x}(u,v)$ by definition, we get the required form

$$\mathbf{x}(u,v) = \operatorname{Re}\left[\alpha(z) - i \int_{u_0}^{z} N(w) \times \alpha'(w)\,dw\right].$$

To prove that a solution to Björling's problem exists, we define a holomorphic curve by $\beta(z) = \left[\alpha(z) - i \int_{u_0}^{z} N(w) \times \alpha'(w)\,dw\right]$ on a domain D where the power series converge. Since $N(z)$ and $\alpha'(z)$ are real for $z \in I$, we have $\operatorname{Re}\beta'(z) = \alpha'(z)$ and $\operatorname{Im}\beta'(z) = -N(z) \times \alpha'(z)$. For $z \in I$, we have $\alpha' \cdot (N \times \alpha') = 0$ and $|N \times \alpha'| = |N| \cdot |\alpha'| = |\alpha'|$

(since $\mathcal{N} \cdot \alpha' = 0$). Then, for $z \in I$ we have

$$
\begin{aligned}
(\beta')^2 &= \alpha'(z) \cdot \alpha'(z) - 2i\alpha'(z) \cdot (\mathcal{N}(z) \times \alpha'(z)) \\
&\quad - (\mathcal{N}(z) \times \alpha'(z)) \cdot (\mathcal{N}(z) \times \alpha'(z)) \\
&= |\alpha'|^2 - 0 - |\mathcal{N} \times \alpha'|^2 \\
&= 0.
\end{aligned}
$$

Again, the Identity theorem (Theorem 2.4.18) implies that the holomorphic function $(\beta')^2$ must be zero on all of D. But this is exactly the situation which led to the Weierstrass-Enneper representations where ϕ was used instead of β'. Hence, we know by Theorem 3.6.2 that the real part of $\int \beta' \, dw = \beta$ is a minimal surface in isothermal parameters. That is, we have a minimal surface defined by

$$
\mathbf{x}(u,v) = \operatorname{Re} \left[\alpha(z) - i \int_{u_0}^{z} \mathcal{N}(w) \times \alpha'(w) \, dw \right].
$$

We must still check the conditions of the problem. Since $\mathcal{N}(u)$ and $\alpha'(u)$ are real for $u \in I$, $\mathbf{x}(u,0) = \operatorname{Re} \beta(u) = \alpha(u)$ and the first condition is satisfied. Also, for $u \in I$, we may compute $\beta'(u)$ two ways; from the definition of $\beta(z)$ and from the harmonic conjugate process used in the proof of uniqueness. We get

$$
\alpha'(u) - i \left(\mathcal{N}(u) \times \alpha'(u) \right) = \beta'(u) = \mathbf{x}_u(u,0) - i \, \mathbf{x}_v(u,0).
$$

Equating real and imaginary parts gives

$$
\mathbf{x}_u(u,0) = \alpha'(u) \qquad \text{and} \qquad \mathbf{x}_v(u,0) = \mathcal{N}(u) \times \alpha'(u).
$$

But isothermal coordinates give

$$
\mathbf{x}_v(u,0) = U(u,0) \times \mathbf{x}_u(u,0) = U(u,0) \times \alpha'(u),
$$

and, consequently, $\mathcal{N}(u) = U(u,0)$. Thus, the second condition of Björling is satisfied also. $\square$

Example 3.9.3 (Geodesics). Assume α is parametrized by arclength. A unit speed curve $\alpha(s)$ has a unit tangent vector $T = \alpha'(s)$ and a unit normal $N = T'/|T'| = \alpha''/|\alpha''|$. The length $|T'|$ is denoted by κ and is called the curvature of the curve. The *binormal* B of α is given by $B = T \times N$. The curve α is determined by (initial conditions and) the system of differential equations called the *Frenet formulas* relating the rates of change of T, N and B (see [**Opr97**]). The vector fields N and B are examples of the $\mathcal{N}$ of Björling's problem.

A unit speed curve in a surface is a *geodesic* if its acceleration is always parallel to the unit normal of the surface. This condition is saying that the acceleration isn't detectable in any tangent plane, so the curve is a surface analogue of a line in $\mathbb{R}^3$. This is so since, in $\mathbb{R}^3$, straight lines are geodesics because they have no acceleration at all. On spheres, geodesics are great circles because the unit normals of great circles point radially. If we choose $\mathcal{N} = N$, the unit normal of a curve α, then, since the direction of the acceleration is precisely the direction of N, we see that the condition of Björling's problem is that the given curve be a geodesic in the surface.

Exercise 3.9.4. For the situation where $\mathcal{N} = N$, the unit normal of α, show that the solution to Björling's problem may be written

$$\mathbf{x}(u,v) = \mathrm{Re}\left[\alpha(z) + i \int_{u_0}^z B|\alpha'(w)|\, dw \right],$$

where $B = T \times N$ is the binormal of α. Hint: $N \times \alpha' = -|\alpha'|(T \times N)$.

Now choose the vector field $-N(t) = \mathcal{N}$.

Corollary 3.9.5. *Let* $\mathbf{x}(u,v)$ *be a solution to Björling's problem with curve* $\alpha(t)$ *and vector field* $N(t)$. *Then, for the curve* $\alpha(t)$ *and vector field* $-N(t)$, *the solution* $\tilde{\mathbf{x}}(u,v)$ *to Björling's problem is given by*

$$\tilde{\mathbf{x}}(u,v) = \mathrm{Re}\left[\alpha(z) + i \int_{u_0}^z N(w) \times \alpha'(w)\, dw \right]$$
$$= \mathbf{x}(u,-v).$$

Proof. Flip the domain D about the u-axis to get a domain $\tilde{D}$ and define $\tilde{\mathbf{x}}(u,v) = \mathbf{x}(u,-v)$. Clearly, this surface is minimal and $\tilde{U}(u,v) = -U(u,-v)$. Hence $\tilde{\mathbf{x}}(u,v)$ solves Björling's problem with vector field $-N(t)$. By the uniqueness of such a solution, however, we also have

$$\tilde{\mathbf{x}}(u,v) = \mathrm{Re}\left[\alpha(z) + i \int_{u_0}^z N(w) \times \alpha'(w)\, dw \right],$$

and we are done. $\qquad\square$

Remark 3.9.6. For $u \in I$ (i.e. $v = 0$), $N(u)$, $\alpha'(u)$ and $z = u$ are real, so $\tilde{\mathbf{x}}(u, 0) = \alpha(u) = \mathbf{x}(u, 0)$. Since D and $\tilde{D}$ are open and contain I, they must overlap in an open set. The Identity theorem (Theorem 2.4.18) applies to show that, in fact, $\mathbf{x}(u, v)$ and $\tilde{\mathbf{x}}(u, v)$ are the same surface on the overlap. This means that we have extended $\mathbf{x}(u, v)$ past its original domain D by piecing it together with $\tilde{\mathbf{x}}(u, v)$ in $\tilde{D}$. This is an example of why the Identity theorem is also called The Principle of Analytic Continuation.

In the proof of the next result, we choose $\mathcal{N}$ to be another vector field which is automatically perpendicular to the tangent vector of a curve in a surface. Namely, we take $\mathcal{N}$ to be the surface normal U itself.

Lemma 3.9.7. *Suppose M is a minimal surface parametrized by $\mathbf{x}(u, v)$ and $\mathbf{x}(u, 0)$ is contained in the xy-plane. If $\mathbf{x}(u, v)$ meets the xy-plane orthogonally along $\mathbf{x}(u, 0)$, then $x^1(u, -v) = x^1(u, v)$, $x^2(u, -v) = x^2(u, v)$ and $x^3(u, -v) = -x^3(u, v)$.*

Proof. Denote $\mathbf{x}(u, 0)$ by $\alpha(u)$ and let the vector field $\mathcal{N}(u)$ along α be defined to be the unit normal of the minimal surface $U(u, 0)$. Since α is in the xy-plane, it may be written $\alpha(u) = (\alpha^1(u), \alpha^2(u), 0)$. Also, since the surface meets the xy-plane orthogonally, $\mathcal{N}(u) = U(u, 0) = (\mathcal{N}^1(u), \mathcal{N}^2(u), 0)$. With a view to applying the solution to Björling's problem, we calculate

$$\mathcal{N}(u) \times \alpha'(u) = (0, 0, \mathcal{N}^1(u)(\alpha^2)'(u) - \mathcal{N}^2(u)(\alpha^1)'(u)).$$

The Identity theorem implies that the holomorphic extensions of these quantities must have exactly the same formulas, but with variable z instead of u. Now we may apply the uniqueness of the solution to Björling's problem. Since the first and second coordinates of $\mathcal{N}(z) \times \alpha'(z)$ are zero, we have $x^1(u, v) = \operatorname{Re} \alpha^1(z)$ and $x^2(u, v) = \operatorname{Re} \alpha^2(z)$. Using Corollary 3.9.5 with α and $-\mathcal{N}$, however, we see that $x^1(u, -v) = \operatorname{Re} \alpha^1(z)$ and $x^2(u, -v) = \operatorname{Re} \alpha^2(z)$ as well. Hence, the first two relations hold. Since the third coordinate of α is zero, $x^3(u, v) = -\operatorname{Re} i \int \mathcal{N} \times \alpha' \, dw$. Again using Corollary 3.9.5 with α and $-\mathcal{N}$, we see that $x^3(u, -v) = +\operatorname{Re} i \int \mathcal{N} \times \alpha' \, dw = -x^3(u, v)$. $\quad\square$

Exercise 3.9.8. Suppose M is a minimal surface parametrized by $\mathbf{x}(u,v)$, and $\mathbf{x}(u,0)$ lies on the x-axis. Show that $x^1(u,-v) = x^1(u,v)$, $x^2(u,-v) = -x^2(u,v)$ and $x^3(u,-v) = -x^3(u,v)$. Hint: $\alpha(u) = (\alpha^1(u), 0, 0)$ and $\mathcal{N}(u) = (0, \mathcal{N}^2(u), \mathcal{N}^3(u))$.

Recall that, given a point p and line l on the surface M, we say that p is *symmetric to a point* $q \in M$ *with respect to* l if q lies the same distance away from l as p and the line in $\mathbb{R}^3$ joining p and q meets l at a right angle. Similarly, given a point p and a plane P meeting the surface at a right angle, p is *symmetric to q with respect to P* if q lies the same distance from P as p and the line joining p and q meets P at a right angle. Lemma 3.9.7 and Exercise 3.9.8, together with our ability to rigidly move surfaces in $\mathbb{R}^3$ without changing their geometry, now prove the following famous result.

Theorem 3.9.9 (The Schwarz Reflection Principles).

(1) *A minimal surface is symmetric about any straight line contained in the surface.*

(2) *A minimal surface is symmetric about any plane which intersects the surface orthogonally.*

Example 3.9.10. Consider Scherk's (first) surface

$$z = c \ln \left(\frac{\cos(x/c)}{\cos(y/c)} \right)$$

on the square $-\pi/2 < x/c < \pi/2$, $-\pi/2 < y/c < \pi/2$. As we approach any of the vertices of the square, we approach an indeterminate form $0/0$ which can approach any real number. That is, extending Scherk's surface to the vertices of the square requires erecting a vertical line over each vertex. But then we may apply symmetry about a line to continue Scherk's surface to a diagonally adjacent square such as $\pi/2 < x/c < 3\pi/2$, $\pi/2 < y/c < 3\pi/2$. Continuing in this manner produces the entire Scherk's surface.

Also notice that, on the square $-\pi/2 < x/c < \pi/2$, $-\pi/2 < y/c < \pi/2$, Scherk's surface intersects the xy-plane in the lines $y = \pm x$. This is easy to see since $z = 0$ when the natural log vanishes; namely, at 1. Hence, $z = 0$ when $\cos(x/c) = \cos(y/c)$. In the square, this means

$y = \pm x$. These lines are lines of symmetry for Scherk's surface. To see this, note that the reflection through $y = x$ (say) in the plane is given by $(a, b) \mapsto (b, a)$ and that

$$\ln\left(\frac{\cos(b/c)}{\cos(a/c)}\right) = -\ln\left(\frac{\cos(a/c)}{\cos(b/c)}\right).$$

Exercise 3.9.11. Show that $y = -x$ is also a line of symmetry for Scherk's surface.

Exercise 3.9.12. Verify the first statement of Theorem 3.9.9 for a ruling $(0, 0, v_0) + u\,(\cos v_0, \sin v_0, 0)$ of the particular helicoid $\mathbf{x}(u, v) = (u\cos v, u\sin v, v)$.

Example 3.9.13 (Plateau's $90°$ Rule Revisited).
Suppose α is a plane curve with unit normal N. The minimal surface which solves Björling's problem for this data has unit normal U equal to N along α. Assuming α has unit speed, we have unit tangent vector $\alpha' = T$ and unit normal N obeying $\kappa N = T' = \alpha''$. Because α lies in a plane P, for every parameter s, it must be the case that $(\alpha(s) - \alpha(0)) \cdot n = 0$, where n is the constant unit normal of P. Differentiating twice, we obtain

$$T \cdot n = 0 \qquad \text{and} \qquad N \cdot n = 0.$$

Then the vectors T and N are both in the plane P also. But if N is in the plane and $N = U$, then $U \cdot n = 0$ too. This means that the minimal surface obtained from solving Björling's problem and the plane P are perpendicular. Furthermore, because the solution to Björling's problem is unique, this is always the case for minimal surfaces intersecting planes (where the intersection curve is allowed to move freely on the plane). This proves Plateau's $90°$ rule (see § 1.5).

Example 3.9.14. If $\alpha(u) = (\beta(u), 0, \gamma(u))$ is in the xz-plane with unit normal vector field $N(u)$, then the solution to Björling's problem is given by

$$\mathbf{x}(u, v) = \left(\operatorname{Re}\beta(z),\ \operatorname{Im}\int\sqrt{\beta'^2(w) + \gamma'^2(w)}\,dw,\ \operatorname{Re}\gamma(z)\right).$$

To see this, note that, for $\alpha' = (\beta', 0, \gamma')$, the unit normal is

$$N = \frac{(-\gamma', 0, \beta')}{\sqrt{\beta^2 + \gamma'^2}}.$$

Then $N \times \alpha' = (0, \sqrt{\beta^2 + \gamma'^2}, 0)$, and $-i\, N \times \alpha'$ has real part the imaginary part of $N \times \alpha'$. Hence,

$$\mathrm{Re}\left(\alpha - i \int N \times \alpha'\right) = \left(\mathrm{Re}\,\beta, \mathrm{Im} \int \sqrt{\beta^2 + \gamma'^2}\, dw, \mathrm{Re}\,\gamma\right).$$

In § 5.10, we shall use Maple to create minimal surfaces from plane curves using this form of the solution to Björling's problem.

Example 3.9.15 (Catalan's Surface). Let

$$\alpha(u) = (1 - \cos u, 0, u - \sin u)$$

be a cycloid. We will find the minimal surface containing α as a geodesic. This surface is Catalan's surface. By Example 3.9.14, we easily have, for $z = u + iv$,

$$x^1(u, v) = \mathrm{Re}(1 - \cos z) \qquad x^3(u, v) = \mathrm{Re}(z - \sin z)$$
$$= 1 - \cos(u)\cosh(v), \qquad\qquad = u - \sin(u)\cosh(v),$$

using Exercise 2.4.1. For $x^2(u, v)$, we must compute $\alpha'(z)$ and the square root of the formula. First, $\alpha' = (\sin z, 0, 1 - \cos z)$. Then

$$\sin^2(w) + (1 - \cos(w))^2 = 2 - 2\cos(w)$$
$$= 4\left(\frac{1 - \cos(w)}{2}\right)$$
$$= 4\sin^2(w/2).$$

Therefore, $\sqrt{\sin^2(w) + (1 - \cos(w))^2} = 2\sin(w/2)$ and

$$\int \sqrt{\sin^2(w) + (1 - \cos(w))^2}\, dw = \int 2\sin(w/2)\, dw = -4\cos(w/2).$$

Again using Exercise 2.4.1, we see that we have $\mathrm{Im}(-4\cos(w/2)) = 4\sin(u/2)\sinh(v/2)$ and, finally,

$$\mathbf{x}(u, v) = \left(1 - \cos(u)\cosh(v),\ 4\sin\frac{u}{2}\sinh\frac{v}{2},\ u - \sin(u)\cosh(v)\right).$$

Exercise 3.9.16 (Henneberg's Surface). A parametrization for Henneberg's surface may be given by

$$\mathbf{x}(u,v) = (x^1(u,v), x^2(u,v), x^3(u,v)),$$

where

$$x^1(u,v) = -1 + \cosh 2u \cos 2v,$$

$$x^2(u,v) = \sinh u \sin v + \frac{1}{3}\sinh 3u \sin 3v,$$

$$x^3(u,v) = -\sinh u \cos v + \frac{1}{3}\sinh 3u \cos 3v.$$

Take a parametrization for Neil's parabola $2x^3 = 9z^2$ given by

$$\alpha(u) = (\cosh 2u - 1, 0, -\sinh u + \frac{1}{3}\sinh 3u).$$

To see that this works, verify the identities $\cosh 2u = 1 + 2\sinh^2 u$ and $\frac{1}{3}\sinh 3u - \sinh u = \frac{4}{3}\sinh^3 u$. Now choose α's unit normal $N(u)$ as the vector field $\mathcal{N}$ in Björling's problem so that α is a geodesic in the resulting surface. Show that

$$x^1(u,v) = -1 + \cosh 2u \cos 2v,$$

$$x^3(u,v) = -\sinh u \cos v + \frac{1}{3}\sinh 3u \cos 3v.$$

For $x^2(u,v)$, compute $\alpha'(z) = (2\sinh 2z, 0, -\cosh z + \cosh 3z)$ and use Example 3.9.14. Calculate

$$\text{Im} \int \sqrt{\beta'^2(w) + \gamma'^2(w)}\, dw =$$

$$\text{Im} \int \sqrt{4\cosh^2 2w - 4 + \cosh^2 w - 2\cosh w \cosh 3w + \cosh^2 3w}\, dw.$$

Show that the integral reduces to

$$\int \sinh 3w + \sinh w\, dw = \frac{1}{3}\cosh 3z + \cosh z,$$

using the identities: $\cosh^2 2z = 4\cosh^4 z - 4\cosh^2 z + 1$, $\cosh 3z = 4\cosh^3 z - 3\cosh z$, $\cosh z \cosh 3z = 4\cosh^4 z - 3\cosh^2 z$, $\cosh^2 3z = 16\cosh^6 z - 24\cosh^4 z + 9\cosh^2 z$ and $\sinh 3z = 3\sinh z + 4\sinh^3 z$.

Now show that

$$\mathrm{Im} \int \sqrt{\beta'^2(w) + \gamma'^2(w)}\, dw = \frac{1}{3}\sinh 3u \sin 3v + \sinh u \sin v.$$

Finally, take the strip in the domain of u and v defined by $u = 2r - 1 + s$, $v = \pi(r - \frac{1}{4})$ with $0 \le r \le 1$, $-.1 \le s \le .1$. Graph the surface obtained by applying the Henneberg parametrization $\mathbf{x}(u,v)$ to this strip. This is a Moebius strip, and its existence inside of Henneberg's surface says that the surface is not *orientable*.

Exercise 3.9.17. Give another proof of the fact that the only minimal surface of revolution is a catenoid. Assume that, for a parametrization of M, $\mathbf{x}(u,v) = (h(u)\cos v, u, h(u)\sin v)$, h has a minimum at $u_0 = 0$, and apply the solution to Björling's problem to the parallel circle at $u_0 = 0$, $\alpha(v) = (h(0)\cos v, 0, h(0)\sin v)$ with vector field the unit normal of the circle. (The assumption $h'(0) = 0$ assures that U is radial on the circle just as N is.)

3.10. To Be or Not To Be Area Minimizing

Minimal surfaces do not always minimize area. We have already seen this for certain catenoids. In this section, we present an approach due to Schwarz which tells us when we have minimal non-area-minimizing surfaces ([**Rad71**]). Here we shall see that all of the work we did with complex variables really pays off, for it allows us to analyze the difficult question of when minimal and area-minimizing mean different things. Before we do this, however, let's look at one simple example where we *can* compare areas using the usual tools of vector analysis.

Example 3.10.1. Suppose $z = f(x,y)$ is a function of two variables *which satisfies the minimal surface equation* (see Proposition 3.2.3). Let the graph of f be parametrized by $\mathbf{x} = (x, y, f(x,y))$ over a closed disk (say) in the xy-plane with boundary curve C. Take any other function $z = g(x,y)$ on the disk with $g|_C = f|_C$, and suppose for convenience that the union of the two graphs along the common boundary C forms a surface with no self-intersections. That is, if $\mathbf{y} = (x, y, g(x,y))$ is a parametrization for the graph of g, then the surface $S = \mathbf{x} \cup_C \mathbf{y}$ encloses a volume in $\mathbb{R}^3$. Now let U be the unit normal vector field for $\mathbf{x}$:

$$U = \frac{(-f_x, -f_y, 1)}{\sqrt{1 + f_x^2 + f_y^2}}.$$

Although we usually think of U as a vector field on $\mathbf{x}$, in fact, there is no reason to not consider it at every point in $\mathbb{R}^3$ above the disk where f is defined. Of course, note that U does not depend on z though. If, without loss of generality, we assume the graph of f lies above that of g, then U is the outer normal of $\mathbf{x}$ and is pointed inside S on $\mathbf{y}$. Now let's compute the divergence of the vector field U:

$$
\begin{aligned}
\mathrm{div}(U) &= -\frac{\partial}{\partial x}\left(\frac{f_x}{\sqrt{1+f_x^2+f_y^2}}\right) - \frac{\partial}{\partial y}\left(\frac{f_y}{\sqrt{1+f_x^2+f_y^2}}\right) \\
&\quad + \frac{\partial}{\partial z}\left(\frac{1}{\sqrt{1+f_x^2+f_y^2}}\right) \\
&= -\frac{f_{xx}(1+f_x^2+f_y^2) - f_x^2 f_{xx} - f_x f_y f_{xy}}{(1+f_x^2+f_y^2)^{3/2}} \\
&\quad - \frac{f_{yy}(1+f_x^2+f_y^2) - f_y f_x f_{xy} - f_y^2 f_{yy}}{(1+f_x^2+f_y^2)^{3/2}} \\
&= -\frac{f_{xx}(1+f_y^2) - 2f_x f_y f_{xy} + f_{yy}(1+f_x^2)}{(1+f_x^2+f_y^2)^{3/2}} \\
&= 0
\end{aligned}
$$

since f satisfies the minimal surface equation. Recall that the divergence theorem says that,

$$\iiint_\Omega \mathrm{div}\, V\, d\Omega = \iint_M V \cdot U\, dA$$

for a surface M enclosing a volume Ω. Here, the divergence of a vector field $V = (V^1, V^2, V^3)$ is defined to be $\mathrm{div}\, V = \partial V^1/\partial x^1 + \partial V^2/\partial x^2 + \partial V^3/\partial x^3$, and U is the unit outward normal of M. Now, using the directions of the vector field U mentioned above and the downward unit normal $\mathcal{N}$ of $\mathbf{y}$ (corresponding to the outer normal on

S), the divergence theorem, together with the calculation of div(U) above, gives

$$0 = \iiint \mathrm{div}(U)\, dx\, dy\, dz$$

$$= \iint_S U \cdot dA_S$$

$$= \iint_{\mathbf{x}} U \cdot dA_{\mathbf{x}} + \iint_{\mathbf{y}} U \cdot dA_{\mathbf{y}}$$

$$= \iint_{\mathbf{x}} U \cdot U \sqrt{1 + f_x^2 + f_y^2}\, dx\, dy + \iint_{\mathbf{y}} U \cdot \mathcal{N} \sqrt{1 + g_x^2 + g_y^2}\, dx\, dy$$

$$= \iint_{\mathbf{x}} \sqrt{1 + f_x^2 + f_y^2}\, dx\, dy + \iint_{\mathbf{y}} \cos(\theta) \sqrt{1 + g_x^2 + g_y^2}\, dx\, dy,$$

where θ is the angle between the unit vectors U and $\mathcal{N}$. But then we have

$$\mathrm{Area}(\mathbf{x}) = \left| \iint \sqrt{1 + f_x^2 + f_y^2}\, dx\, dy \right|$$

$$= \left| \iint \cos(\theta) \sqrt{1 + g_x^2 + g_y^2}\, dx\, dy \right|$$

$$\leq \iint |\cos(\theta)| \sqrt{1 + g_x^2 + g_y^2}\, dx\, dy$$

$$\leq \iint \sqrt{1 + g_x^2 + g_y^2}\, dx\, dy$$

$$= \mathrm{Area}(\mathbf{y}).$$

So, satisfying the minimal surface equation *can* guarantee minimum surface area when compared to the right other surfaces. A more general version of this result together with a proof of a different caliber may be found in [**Mor88**, §6.1].

Now let's consider the general situation. Take a minimal surface M bounded by a curve C and, by Theorem 3.4.1, suppose that the parametrization $\mathbf{x}(u, v)$ of M is isothermal. Hence, all the consequences of having isothermal parameters follow: $E = G$, $F = 0$, $\ell = -n$ (by Exercise 3.4.3), $K = -(\ell^2 + m^2)/E^2$ (see Exercise 3.7.7), $U_u \cdot U_u = U_v \cdot U_v = (\ell^2 + m^2)/E$, $U_u \cdot U_v = 0$ (see the proof of Proposition 3.7.6). Also, of course, we can suppose the isothermal parameters come from a Weierstrass-Enneper representation II associated to a holomorphic function F. From Theorem 3.6.19, we have

$$K = \frac{-4}{|F|^2(1 + u^2 + v^2)^4} \quad \text{and} \quad E = |F|^2(1 + u^2 + v^2)^2 = G.$$

Now take a variation $\mathbf{y}^t(u, v) = \mathbf{x}(u, v) + tV(u, v)$, where $V(u, v) = \rho(u, v)\,U(u, v)$ is a normal vector field on M of varying length $\rho(u, v)$ with $\rho(C) = 0$. (Here we write $\rho(C) = 0$ to mean that ρ vanishes on the curve in the uv-plane which is carried to C by $\mathbf{x}(u, v)$.) We calculate $\mathbf{y}_u^t = \mathbf{x}_u + tV_u$, $\mathbf{y}_v^t = \mathbf{x}_v + tV_v$ and $\mathbf{y}_u^t \times \mathbf{y}_v^t = \mathbf{x}_u \times \mathbf{x}_v + t[\mathbf{x}_u \times V_v + V_u \times \mathbf{x}_v] + t^2 V_u \times V_v$. In the following, we use the notation

$$\maltese = (\mathbf{x}_u \times \mathbf{x}_v) \cdot (\mathbf{x}_u \times V_v + V_u \times \mathbf{x}_v),$$

$$\maltese \diamond \maltese = 2(\mathbf{x}_u \times \mathbf{x}_v) \cdot (V_u \times V_v) + (\mathbf{x}_u \times V_v) \cdot (\mathbf{x}_u \times V_v)$$
$$+ 2(\mathbf{x}_u \times V_v) \cdot (V_u \times \mathbf{x}_v) + (V_u \times \mathbf{x}_v) \cdot (V_u \times \mathbf{x}_v),$$

$$\mathcal{S} = \sqrt{|\mathbf{x}_u \times \mathbf{x}_v|^2 + 2t\maltese + t^2\maltese\diamond\maltese + O(t^3)},$$

where $O(t^3)$ denotes terms involving powers of t greater than or equal to three. With this notation, we see that the surface area $A(t) = \iint |\mathbf{y}_u^t \times \mathbf{y}_v^t|\, du\, dv$ is given by $A(t) = \iint \mathcal{S}\, du\, dv$. We assume that $t = 0$ (corresponding to M) is a critical point for the area, so $A'(0) = 0$. We shall use the result of the following simple exercise extensively below.

Exercise 3.10.2 (Lagrange identity). Show that the following is always true (Maple does this quickly):

$$(\mathbf{v} \times \mathbf{w}) \cdot (\mathbf{a} \times \mathbf{b}) = (\mathbf{v} \cdot \mathbf{a})(\mathbf{w} \cdot \mathbf{b}) - (\mathbf{v} \cdot \mathbf{b})(\mathbf{w} \cdot \mathbf{a}).$$

Lemma 3.10.3. *If M is minimal, then $\maltese = 0$.*

Proof. First note that

$$\maltese = (\mathbf{x}_u \times \mathbf{x}_v) \cdot (\mathbf{x}_u \times V_v) + (\mathbf{x}_u \times \mathbf{x}_v) \cdot (V_u \times \mathbf{x}_v)$$
$$= E(\mathbf{x}_v \cdot V_v) + E(\mathbf{x}_u \cdot V_u)$$

by Exercise 3.10.2 using isothermal parameters. Now, $V_u = \rho_u U + \rho U_u$ and $V_v = \rho_v U + \rho U_v$, so we have

$$\begin{aligned}
\mathbf{x}_u \cdot V_u &= \mathbf{x}_u \cdot (\rho_u U + \rho U_u)\\
&= \rho \mathbf{x}_u \cdot U_u\\
&= -\rho \ell
\end{aligned}$$

by definition of ℓ. Similarly, $\mathbf{x}_u \cdot V_v = -\rho m$, $\mathbf{x}_v \cdot V_u = -\rho m$ and $\mathbf{x}_v \cdot V_v = -\rho n = \rho \ell$. We then have

$$\maltese = \rho E \ell - \rho E \ell = 0.$$

$\square$

Now let's look at $\maltese\!\bullet\!\maltese$ using Exercise 3.10.2 and the calculations above. Note that $V_u = \rho_u U + \rho U_u$, $V_v = \rho_v U + \rho U_v$, $U_u \cdot U_u = U_v \cdot U_v = (\ell^2 + m^2)/E$ and $U_u \cdot U_v = 0$. We have

$$\begin{aligned}
\maltese\!\bullet\!\maltese &= 2(\mathbf{x}_u \times \mathbf{x}_v) \cdot (V_u \times V_v) + (\mathbf{x}_u \times V_v) \cdot (\mathbf{x}_u \times V_v)\\
&\quad + 2(\mathbf{x}_u \times V_v) \cdot (V_u \times \mathbf{x}_v) + (V_u \times \mathbf{x}_v) \cdot (V_u \times \mathbf{x}_v)\\
&= 2[(\mathbf{x}_u \cdot V_u)(\mathbf{x}_v \cdot V_v) - (\mathbf{x}_u \cdot V_v)(\mathbf{x}_v \cdot V_u)]\\
&\quad + (\mathbf{x}_u \cdot \mathbf{x}_u)(V_v \cdot V_v) - (\mathbf{x}_u \cdot V_v)(V_v \cdot \mathbf{x}_u) + 2[(\mathbf{x}_u \cdot V_u)(V_v \cdot \mathbf{x}_v)\\
&\quad - (\mathbf{x}_u \cdot \mathbf{x}_v)(V_v \cdot V_u)] + (V_u \cdot V_u)(\mathbf{x}_v \cdot \mathbf{x}_v) - (V_u \cdot \mathbf{x}_v)(\mathbf{x}_v \cdot V_u)\\
&= 2[(-\rho\ell)(\rho\ell) - (-\rho m)(-\rho m)] + E(\rho_v^2 + \rho^2(\ell^2 + m^2)/E)\\
&\quad - \rho^2 m^2 + 2(-\rho\ell)(\rho\ell) + E(\rho_u^2 + \rho^2(\ell^2 + m^2)/E) - \rho^2 m^2\\
&= 4\rho^2[-\ell^2 - m^2] + \rho^2[\ell^2 + m^2 + \ell^2 + m^2] + E(\rho_v^2 + \rho_u^2)\\
&= 2\rho^2[-\ell^2 - m^2] + E(\rho_v^2 + \rho_u^2)\\
&= 2\rho^2 E^2 K + E(\rho_u^2 + \rho_v^2).
\end{aligned}$$

So S simplifies to $S = \sqrt{|\mathbf{x}_u \times \mathbf{x}_v|^2 + t^2 \maltese\!\bullet\!\maltese + O(t^3)}$, and we differentiate $A(t)$ to obtain

$$A'(t) = \iint \frac{t(2\rho^2 E^2 K + E(\rho_u^2 + \rho_v^2)) + O(t^2)}{S} \, du \, dv$$

and

$$A''(t) = \iint \frac{(2\rho^2 E^2 K + E(\rho_u^2 + \rho_v^2) + O(t))S - t(2\rho^2 E^2 K + E(\rho_u^2 + \rho_v^2))S'}{S^2} \, du \, dv.$$

Note that $S' = \frac{t \maltese \maltese + O(t^2)}{S}$ by Lemma 3.10.3, so $S|_{t=0} = |\mathbf{x}_u \times \mathbf{x}_v| = \sqrt{E^2} = E$ and $S'|_{t=0} = 0$. Therefore, we have

$$A''(0) = \iint \frac{2\rho^2 E^2 K + E(\rho_u^2 + \rho_v^2)}{E} \, du \, dv$$

$$= \iint 2\rho^2 EK + \rho_u^2 + \rho_v^2 \, du \, dv.$$

Substituting the formulas for K and E derived from the Weierstrass-Enneper representation into $A''(0)$ gives

$$A''(0) = \iint \frac{(2\rho^2|F|^2(1+u^2+v^2)^2)(-4)}{|F|^2(1+u^2+v^2)^4} + \rho_u^2 + \rho_v^2 \, du \, dv$$

$$= \iint \frac{-8\rho^2}{(1+u^2+v^2)^2} + \rho_u^2 + \rho_v^2 \, du \, dv.$$

The integration to calculate $A''(0)$ is carried out over a region $\mathcal{R}$ in the uv-parameter plane, and the expression for $A''(0)$ does not depend on the Weierstrass-Enneper representation at all, but only on $\mathcal{R}$ and the choice of ρ on that region. As in ordinary calculus, if $t = 0$ gives a minimum, then the second derivative must be non-negative there. Hence, if we can find a function ρ (with $\rho(C) = 0$) defined on $\mathcal{R}$ such that $A''(0) < 0$, then the minimal surface M cannot have minimum area among surfaces spanning C. Therefore, we have

Theorem 3.10.4 (Schwarz). *Let M be a minimal surface spanning a curve C. If the closed unit disk $D = \{(u,v) | u^2 + v^2 \leq 1\}$ is contained in the interior of $\mathcal{R}$, then a function ρ exists for which $A''(0) < 0$. Hence, M does not have minimum area among surfaces spanning C.*

Proof. Let $\mathcal{D} = \{(u, v, r) \mid u^2 + v^2 \leq r^2\}$ be the domain bounded by the cone $r = \sqrt{u^2 + v^2}$. Define a function on $\mathcal{D}$ by

$$\rho(u, v, r) = \frac{u^2 + v^2 - r^2}{u^2 + v^2 + r^2}$$

and consider

$$\mathcal{A}(r) \overset{\text{def}}{=} \iint_{D(r)} \frac{-8\rho^2}{(1+u^2+v^2)^2} + \rho_u^2 + \rho_v^2 \, du \, dv$$

where $D(r) = \{(u,v) | u^2 + v^2 < r^2\}$ is the open r-disk. Of course, this is $A''(0)$ when we are in $\mathcal{R}$, so we wish to show that the choice of ρ

above leads to $\mathcal{A}(r) < 0$ for certain values of r. First, let's split the integral into two pieces and look at the integral of the last two terms of the integrand. If we let $P = -\rho\rho_v$, $Q = \rho\rho_u$ and apply Green's theorem, we obtain

$$\int_{u^2+v^2=r^2} -\rho\rho_v \, du + \rho\rho_u \, dv$$

$$= \iint_{D(r)} \rho_u^2 + \rho_v^2 \, du \, dv + \iint_{D(r)} \rho(\rho_{uu} + \rho_{vv}) \, du \, dv.$$

Now, the left-hand side is zero because $\rho(u, v, r) = 0$ whenever $u^2 + v^2 = r^2$, so

$$\iint_{D(r)} \rho_u^2 + \rho_v^2 \, du \, dv = -\iint_{D(r)} \rho \, \Delta\rho \, du \, dv,$$

where $\Delta\rho = \rho_{uu} + \rho_{vv}$ is the Laplacian of ρ. Hence,

$$\mathcal{A}(r) = \iint_{D(r)} -\rho \left[\frac{8\rho^2}{(1 + u^2 + v^2)^2} + \Delta\rho \right] du \, dv.$$

Exercise 3.10.5. Show by direct calculation, for $\rho = \rho(u, v, 1) = \frac{u^2+v^2-1}{u^2+v^2+1}$, that

$$\frac{8\rho^2}{(1 + u^2 + v^2)^2} + \Delta\rho = 0.$$

By the exercise, we see that $\mathcal{A}(1) = 0$. Now, $r = 1$ corresponds to the unit disk, so we would like $\mathcal{A}(r) < 0$ for r's slightly larger than 1. To show this is true, we need to look at $\mathcal{A}'(1)$. For convenience, let us change variables by taking $s = \frac{u}{r}$ and $t = \frac{v}{r}$ with $du = r \, ds$, $dv = r \, dt$. Then

$$\mathcal{A}(r) = -\iint \rho \left[\frac{8\rho^2}{(1 + r^2(s^2 + t^2))^2} + \Delta\rho \right] r^2 \, ds \, dt,$$

$$\rho(s, t, r) = \frac{r^2 s^2 + r^2 t^2 - r^2}{r^2 s^2 + r^2 t^2 + r^2} = \frac{s^2 + t^2 - 1}{s^2 + t^2 + 1} \stackrel{\text{def}}{=} \rho(s, t).$$

Therefore, in st-coordinates, ρ does not depend on r. Also, $\rho_u = \rho_s/r$ and $\rho_{uu} = \rho_{ss}/r^2$ by the chain rule, and similarly for v. Hence,

$\Delta_{u,v}\rho = \Delta_{s,t}\,\rho/r^2$. Now we can carry out multiplication by r^2 and replace the uv-Laplace operator in the integral above to get

$$A(r) = -\iint \rho\left[\frac{8r^2\rho^2}{(1+r^2(s^2+t^2))^2} + \Delta_{s,t}\,\rho\right]ds\,dt.$$

The fact that ρ and $\Delta_{s,t}\,\rho$ are independent of r allows us to easily take the derivative of $A(r)$ with respect to r. We obtain

$$A'(r) = -\iint \frac{16\rho^2 r(1+r^2(s^2+t^2))^2 - 32\rho^2 r^3(1+r^2(s^2+t^2))(s^2+t^2)}{(1+r^2(s^2+t^2))^4}\,ds\,dt.$$

At $r = 1$, we have $s^2+t^2 < 1$, so replacing ρ by its definition in terms of s and t gives

$$\begin{aligned}
A'(1) &= -\iint \frac{16\rho^2(1+s^2+t^2) - 32\rho^2(s^2+t^2)}{(1+s^2+t^2)^3}\,ds\,dt \\
&= -\iint_{s^2+t^2<1} \frac{16\rho^2(1+s^2+t^2-2s^2-2t^2)}{(1+s^2+t^2)^3}\,ds\,dt \\
&= -\iint_{s^2+t^2<1} \frac{16(s^2+t^2-1)^2(1-s^2-t^2)}{(1+s^2+t^2)^5}\,ds\,dt \\
&= -\iint_{s^2+t^2<1} \frac{-16(s^2+t^2-1)^2(s^2+t^2-1)}{(1+s^2+t^2)^5}\,ds\,dt \\
&= 16\iint_{s^2+t^2<1} \frac{(s^2+t^2-1)^3}{(1+s^2+t^2)^5}\,ds\,dt.
\end{aligned}$$

The numerator of the integrand is always negative since $s^2 + t^2 < 1$, so $A'(1) < 0$. Since this means that $A(r)$ is decreasing at $r = 1$ and we have seen previously that $A(1) = 0$, it must be the case that $A(r) < 0$ for $1 < r < \bar{r}$ (some $\bar{r}$).

Now suppose that the unit disk D is contained in the interior of the parameter domain $\mathcal{R}$. Then there is an r such that $1 < r < \bar{r}$ and $\{(u,v)|u^2 + v^2 \le r^2\} \subseteq \mathcal{R}$. Define

$$\rho(u,v) = \begin{cases} \rho(u,v,r) & \text{for } u^2 + v^2 \le r^2, \\ 0 & \text{for } u^2 + v^2 > r^2. \end{cases}$$

Note that $\rho|_{\partial\mathcal{R}} = 0$ and, by the discussion above, $A''(0) < 0$. Note that the partial derivatives ρ_u, ρ_v are not continuous on the boundary circle $\{(u,v)|u^2 + v^2 = r^2\}$, but may be 'rounded off' suitably there while keeping $A''(0) < 0$. Hence, the minimal surface given by the Weierstrass-Enneper representation which spans C is not area minimizing. □

Now let's interpret Theorem 3.10.4 geometrically. Again, the key point here is that the Weierstrass-Enneper representation is not just a bunch of formulas, but a way of obtaining geometric information about the minimal surface from its representation directly. As soon as we described M in terms of the Weierstrass-Enneper representation II, we identified the parameters u and v with the real and imaginary parts of the complex variable τ. Remember that τ is itself identified with the function g of the Weierstrass-Enneper representation I, and g is the Gauss map followed by stereographic projection (Theorem 3.8.1). Therefore, since stereographic projection from the North pole projects the lower hemisphere of S^2 onto the unit disk D, $\mathcal{R}$ contains D in its interior precisely when the image of the Gauss map of M contains the lower hemisphere of S^2 in its interior. There is nothing special about doing stereographic projection from the North pole. Since it can be done from any point on the sphere, we have the following geometrical version of Schwarz's theorem.

Theorem 3.10.6. *Let M be a minimal surface spanning a curve C. If the image of the Gauss map of M contains a hemisphere of S^2 in its interior, then M does not have minimum area among surfaces spanning C.*

Enneper's surface $\mathbf{x}(u, v) = (u - u^3/3 + uv^2, \ -v + v^3/3 - vu^2, \ u^2 - v^2)$ has no self-intersections for $u^2 + v^2 < 3$, and its Gauss map (restricted to the disk $u^2 + v^2 < 3$) covers more than a hemisphere of S^2 (see § 5.8). By Theorem 3.10.6, we infer that Enneper's surface does not minimize area among all surfaces spanning the curve C given by applying the parametrization $\mathbf{x}$ to the parameter circle $u^2 + v^2 = R^2$, where $1 < R < \sqrt{3}$. Of course, by the theorem of Douglas and Radó, there exists a least area (and hence minimal) surface spanning C. Therefore, there are at least two minimal surfaces spanning C. In fact, the question of uniqueness for the solution to Plateau's problem is a delicate one. Here is a positive result along this line which we mention without proof.

Theorem 3.10.7 (Ruchert). *For $0 < r \leq 1$, Enneper's surface is the unique solution to Plateau's problem for the curve given by applying the Enneper parametrization to a parameter circle of radius r.*

Although the result of Douglas and Radó shows that there are at least two minimal surfaces spanning the curve C, we only know one of these explicitly — Enneper's surface. At present, there seems to be no example of two or more explicit minimal surfaces spanning a given curve. Of course, for two curves, there is § 5.6.

3.11. Constant Mean Curvature

Now let's look at situations where mean curvature is non-zero, but constant. The first example which springs to mind is a sphere because, at each point, the normal curvatures are the same in any direction. Therefore, the mean curvature is also this constant normal curvature. Let's verify this intuition using our formulas.

Example 3.11.1 (The Sphere). A parametrization for the R-sphere is given by

$$\mathbf{x}(u,v) = (R\cos(u)\cos(v), R\sin(u)\cos(v), R\sin(v)).$$

This parametrization gives $E = R^2 \cos^2(v)$, $F = 0$ and $G = R^2$ as well as $\ell = -R\cos^2(v)$, $m = 0$ and $n = -R$. Thus,

$$
\begin{aligned}
H &= \frac{G\ell + En - 2Fm}{2(EG - F^2)} \\
&= \frac{R^2(-R\cos^2(v)) + R^2\cos^2(v)(-R) - 0}{2(R^2\cos^2(v)R^2)} \\
&= \frac{-2R^3\cos^2(v)}{2(R^4\cos^2(v))} \\
&= -\frac{1}{R}.
\end{aligned}
$$

Exercise 3.11.2. Take a surface of revolution parametrized by $\mathbf{x}(u,v) = (u, h(u)\cos v, h(u)\sin v)$.

(1) Show that the mean curvature is given by

$$H = \frac{1}{2}\frac{-hh'' + 1 + h'^2}{h(1 + h'^2)^{3/2}}.$$

Suppose throughout the problem that $H = c/2$ is constant, and derive the differential equation

$$1 + h'^2 - hh'' = ch(1 + h'^2)^{3/2}.$$

(2) Let $c = 0$ and show that the differential equation then reduces to $1 + h'^2 - hh'' = 0$. Have you seen this equation before? Solve it if you don't recognize it. What is the condition $c = 0$ saying geometrically?

(3) Now suppose $c = \pm 1/a$ with $a > 0$. Show that the function $h(u)$ satisfies the differential equation

$$h^2 \pm \frac{2ah}{\sqrt{1 + h'^2}} = \pm b^2.$$

(We give the equation in this form to match up with Exercise 4.5.12, but see Exercise 3.11.3 for another formulation.) Surfaces of revolution with constant mean curvature have a generating curve given by this differential equation and are called surfaces of Delaunay. They may also be described as roulettes of conics (see [**Del41**], [**Eel87**] or [**Opr97**]). See Exercise 4.5.12 for a variational approach and § 5.13 for a Maple approach. Also, we mention here that the surfaces of Delaunay even play a role in the proof of the Double Bubble Theorem 1.7.4.

Exercise 3.11.3. Show that, if we have a surface of revolution and H is constant, then

$$\frac{d}{du}\left(Hh^2 - \frac{h}{\sqrt{1 + h'^2}}\right) = 0.$$

Thus, the quantity inside the parentheses is a constant c. What is $h(u)$ when $H = 0$ (see Theorem 3.2.5)? Also, prove that $c = 0$ gives a sphere. Now show that

$$\frac{du}{dh} = \frac{Hh^2 - c}{\sqrt{h^2 - (Hh^2 - c)^2}}.$$

This makes sense only when $h^2 - (Hh^2 - c)^2 > 0$, and solving $h^2 - (Hh^2 - c)^2 = 0$ for h^2 produces another condition: $-1/4 < Hc$. In

§ 5.13, we will see that H and c of the same sign gives an *unduloid*, while H and c of opposite signs (with $-1/4 < Hc < 0$) gives a *nodoid*.

Let M be a compact oriented surface in $\mathbb{R}^3$ with unit normal $U = (\mathbf{x}_u \times \mathbf{x}_v)/|\mathbf{x}_u \times \mathbf{x}_v|$, mean curvature H and area

$$A = \iint |\mathbf{x}_u \times \mathbf{x}_v| \, du \, dv.$$

Just as in the derivation of Theorem 3.3.6, we perturb M a bit by a vector field V on M. Note that we use a vector field here instead of a function g because M is not defined as the graph over some domain. We have

$$M^t\colon \mathbf{y}^t(u,v) = \mathbf{x}(u,v) + tV(u,v)$$

and write the area of M^t as

$$A(t) = \iint |\mathbf{y}_u \times \mathbf{y}_v| \, du \, dv$$
$$= \iint \sqrt{|\mathbf{x}_u \times \mathbf{x}_v|^2 + 2t\,\maltese + O(t^2)} \, du \, dv,$$

where $\maltese = (\mathbf{x}_u \times \mathbf{x}_v) \cdot (\mathbf{x}_u \times V_v + V_u \times \mathbf{x}_v)$ as before. Then, taking the derivative with respect to t and evaluating at $t = 0$, we obtain

$$A'(0) = \iint \frac{\mathbf{x}_u \times \mathbf{x}_v}{|\mathbf{x}_u \times \mathbf{x}_v|} \cdot (\mathbf{x}_u \times V_v + V_u \times \mathbf{x}_v) \, du \, dv$$
$$= \iint \frac{\mathbf{x}_u \times \mathbf{x}_v}{|\mathbf{x}_u \times \mathbf{x}_v|} \cdot (\mathbf{x}_u \times V_v - \mathbf{x}_v \times V_u) \, du \, dv$$
$$= \iint V_v \cdot U \times \mathbf{x}_u - V_u \cdot U \times \mathbf{x}_v \, du \, dv.$$

Let $P = -V \cdot U \times \mathbf{x}_v$ and $Q = V \cdot U \times \mathbf{x}_u$, and calculate

$$\frac{\partial P}{\partial u} = -V_u \cdot U \times \mathbf{x}_v - V \cdot U_u \times \mathbf{x}_v - V \cdot U \times \mathbf{x}_{uv}$$
$$\frac{\partial Q}{\partial v} = V_v \cdot U \times \mathbf{x}_u + V \cdot U_v \times \mathbf{x}_u + V \cdot U \times \mathbf{x}_{uv}.$$

Now apply Green's theorem and Proposition 2.5.1 to get

$$\int_C -P\,dv + Q\,du = \iint \frac{\partial Q}{\partial v} + \frac{\partial P}{\partial u}\,du\,dv$$

$$= \iint V_v \cdot U \times \mathbf{x}_u - V_u \cdot U \times \mathbf{x}_v\,du\,dv$$

$$+ \iint V \cdot (U_v \times \mathbf{x}_u - U_u \times \mathbf{x}_v)du\,dv$$

$$= A'(0) + \iint V \cdot 2H\mathbf{x}_u \times \mathbf{x}_v\,du\,dv.$$

When we integrate over the entire collection of parametrizations on M, the line integrals around the boundaries all cancel because they are taken two ways due to the orientation of M. Hence, the left-hand side above is zero when the integrals are taken over all of M, and we have

(*) $$A'(0) = -\iint 2HU \cdot V\,dA$$

since, by definition, $dA = |\mathbf{x}_u \times \mathbf{x}_v|\,du\,dv$.

Example 3.11.4. Let $V = fU$, a vector field with varying length, but in the direction of the unit normal to M. This is the type of force field obtained from pressure exerted uniformly on a surface. Now, $U \cdot U = 1$, so we have $A'(0) = -2 \iint fH\,dA$. Also, the geometric interpretation of the divergence of a vector field as a rate of change of volume V (see [**MT88**]) together with the divergence theorem gives

$$V'(0) = \iiint \operatorname{div} f U\,dx\,dy\,dz$$

$$= \iint fU \cdot U\,dA$$

$$= \iint f\,dA\,.$$

So, the problem of minimizing surface area (i.e. $A'(0) = 0$ on the surface) subject to having a fixed volume (i.e. $V'(0) = 0$) is equivalent to finding H on an M satisfying

$$\iint fH\,dA = 0 \quad \text{whenever } f \text{ satisfies} \quad \iint f\,dA = 0.$$

This condition implies $H = c$, a constant. To see this, write $H = c + (H - c) = c + J$, where the constant c is chosen to have value $c = \iint H\, dA/A$. Here, A denotes the total area $\iint dA$. The definition of c says that $\iint H - c\, dA = 0$, so $H - c$ will be the function f in the discussion above. Then

$$\iint J^2\, dA = \iint (H - c)(H - c)\, dA$$
$$= \iint H(H - c)\, dA - c\iint H - c\, dA$$
$$= 0,$$

using the condition above to determine H, and also $\iint H - c\, dA = 0$. Then J^2 must be zero, and therefore $J = H - c = 0$ as well. Hence, $H = c$, and we have proven

Theorem 3.11.5. *A closed surface which minimizes surface area subject to a fixed enclosed volume must have constant mean curvature.*

(Also see Example 4.5.10.) Of course, the volume of air inside a bubble is fixed and surface tension works to shrink the bubble's surface area, so this is exactly the situation above. Therefore,

Theorem 3.11.6. *A soap bubble must always take the form of a surface of constant mean curvature.*

But what are these surfaces then? We saw in Example 3.11.1 and in Exercise 3.11.2 that spheres and certain types of surfaces of revolution have constant mean curvature. Of these, however, only spheres are compact (i.e. closed and bounded) and embedded without self-intersections in $\mathbb{R}^3$. So, could it be true that soap bubbles can only be spherical? What did § 1.3.6 lead you to believe? Let's look at the situation mathematically. First, let's derive another consequence of the variational argument above.

Example 3.11.7. Let $V = \mathbf{x}(u, v)$, so that $\mathbf{y}^t = (1+t)\mathbf{x}$ and $A(t) = (1+t)^2 A$ since M is simply stretched uniformly. Clearly, $A'(0) = 2A$, so by $(*)$ we have

$$A = -\iint HU \cdot \mathbf{x}\, dA.$$

Thus we get a formula connecting surface area to mean curvature.

Now we are going to use an estimate due to Antonio Ros to prove a famous theorem of Alexandrov. We will not prove Ros's estimate here, but a proof may be found in [**Opr97**] based on that in [**Oss90**].

Theorem 3.11.8 (Ros's Estimate [**Ros88**]). *Let M be a compact surface embedded in $\mathbb{R}^3$ bounding a domain D of volume V. If $H > 0$ on M, then*

$$\int_M \frac{1}{H}\, dA \geq 3V.$$

Furthermore, equality holds if and only if M is the standard sphere.

Theorem 3.11.9 (Alexandrov). *If M is a compact surface of constant mean curvature embedded in $\mathbb{R}^3$, then M is a standard sphere.*

Proof. Suppose H is constant, and use the area formula of Example 3.11.7 and the divergence theorem to get

$$A = -\iint HU \cdot \mathbf{x}\, dA$$
$$= H \iiint \operatorname{div} \mathbf{x}\, dx\, dy\, dz$$
$$= 3HV.$$

Here we have used the inner normal to M, thus eliminating the minus sign between steps one and two. Then we have

$$\iint \frac{1}{H} dA = \frac{1}{H} \iint dA = \frac{A}{H} = \frac{A}{A/3V} = 3V,$$

and by Ros's estimate, M must be a sphere. $\qquad\square$

Results such as Alexandrov's Theorem led H. Hopf to conjecture that all *immersed* surfaces of constant mean curvature are spheres. An immersed surface may have self-intersections, but still must have the property that $\mathbf{x}_u$ and $\mathbf{x}_v$ are linearly independent at every point. It is only recently that this conjecture has been shown to be false. H. Wente [**Wen86**] was the first to construct tori of constant mean curvature which are immersed in $\mathbb{R}^3$. Since Wente's work, many such examples have been constructed, with important consequences for mechanics and the study of integrable systems. In particular, N. Kapouleas and U. Pinkall-I. Sterling have created and classified such surfaces. For interactions with computer graphics, see [**PS93**]. An example of a surface of constant mean curvature, Sievert's surface, is plotted at the end of § 5.13.

In this chapter, we have seen that the study of soap films and bubbles (i.e. minimal surfaces and surfaces of constant mean curvature) encompasses many different areas and viewpoints in modern mathematics. In the next chapter, we shall more closely examine one of these viewpoints, with an eye toward turning the underlying modes of soap film formation into a general scientific principle.

Chapter 4

The Calculus of Variations and Shape

4.1. Introduction

In Theorem 3.3.6, we saw that least-area surfaces are always minimal. The approach used to prove that result is the fundamental technique of the calculus of variations, a subject whose history begins at the dawn of calculus and whose creators include Bernoulli(s), Euler and Lagrange. It is a subject which has had an impact on not only geometry, but much of science and engineering, including both classical and quantum mechanics. The basic idea behind the application of the calculus of variations to all of these subjects is that things happen because there is some quantity (whether physical or mathematical) which demands to be 'minimized'. In this chapter, we shall see how this demand for economy favors certain types of shapes. Of course, we shall re-prove many things we already know about minimal surfaces, but we also shall expand our study a bit to include other shapes, to indicate exactly how the calculus of variations fits into science and mathematics. Basic references for the subject are [**Tro96**], [**Sag92**] and [**Wei74**] for example. A tremendously readable reference for the history and philosophy of the subject is [**HT85**].

The subject itself can be very technical, so this short chapter can only be considered the briefest of introductions. We will concentrate on the basic techniques and results of the calculus of variations while trying to avoid deeper technical points. Also, we are *only* interested here in how the calculus of variations applies to shape determination, so we will not talk about important topics such as Hamilton's principle. To start, let's take a problem from beginning calculus and see how its solution provides us with a 'shape' as an answer.

Example 4.1.1 (The Swimmer and the Lifeguard). Suppose that a beach is bounded by a straight shoreline and a lifeguard spots a swimmer starting to be pulled under the waves. The lifeguard has to calculate the quickest way to get to the swimmer. She knows that she can run with speed v_1 on the sand and swim with speed v_2. She also knows that a straight line is the shortest distance between two points, so she knows she will run straight to some spot on the shoreline and then swim from there straight to the swimmer. (Here we ignore any currents.) The question really is, what point on the shoreline should she aim for? Of course, we all know that distance equals rate times time, $D = RT$, so the lifeguard's time will be $D_{\text{sand}}/v_1 + D_{\text{water}}/v_2$. To see what the distances are, she estimates that there is a distance L parallel to the shoreline from her position to the swimmer's position. Also, the lifeguard estimates her perpendicular distance to the shoreline to be a and the swimmer's to be b. Then if she runs to a point on the shoreline a parallel distance x from her, the distances she travels are

$$D_{\text{sand}} = \sqrt{a^2 + x^2} \qquad D_{\text{water}} = \sqrt{b^2 + (L-x)^2}.$$

Then the total time for the lifeguard to get to the swimmer is

$$T = \frac{\sqrt{a^2 + x^2}}{v_1} + \frac{\sqrt{b^2 + (L-x)^2}}{v_2}.$$

Having just completed freshman calculus, the lifeguard knows that, in order to find the minimum time, she must take the derivative, set it equal to zero and solve:

$$0 = \frac{dT}{dx}$$

$$0 = \frac{x}{v_1\sqrt{a^2 + x^2}} - \frac{L - x}{v_2\sqrt{b^2 + (L - x)^2}}$$

$$\frac{1}{v_2}\frac{L - x}{\sqrt{b^2 + (L - x)^2}} = \frac{1}{v_1}\frac{x}{\sqrt{a^2 + x^2}}$$

$$\frac{\sin(\theta_w)}{v_2} = \frac{\sin(\theta_s)}{v_1}.$$

Here, θ_s is the angle at x between the lifeguard's path and the perpendicular to the shoreline on the sand side, while θ_w is the angle at x between the lifeguard's (swim) path and the perpendicular to the shoreline. The lifeguard now knows that she should run to the place x on the shore where her path makes an angle θ_s with the perpendicular to the shore.

The reader may recognize the formula above as Snell's law for light refraction when light passes from one medium to another. In fact, Johann Bernoulli used Snell's law to solve the first problem in the calculus of variations, the brachistochrone (see Example 4.4.2). The important point for us to see here is that the final answer is a prescription for the right shape of the solution. (Of course, for this problem, the exact point on the shore may be determined precisely, but we shall see that this is not always the case.) For the swimmer and the lifeguard, then, *the minimum time is attained when the correct geometry obtains.* (The reader can check that the answer above is a true minimum.) This is a paradigm for solving optimization problems using the calculus of variations. The solution of an optimization problem is often described geometrically rather than (or as well as) analytically. Finally, note that we made one very special assumption in the problem: that shortest distances are straight lines. This allowed us to define a single variable x upon which time depended. If we did not know that straight lines give shortest distances, then the simple formulas for distances would instead be arclength integrals over arbitrary paths joining the swimmer and the lifeguard. *Then, among all the paths joining the swimmer and the lifeguard, we would have to find the path which minimized the integrals.* This is an infinitely more difficult problem than finding minima in ordinary calculus, but *this* is the calculus of variations!

4.2. Minimizing Integrals

Let $y = y(x)$ denote a function of x with fixed endpoints $y(x_0) = y_0$ and $y(x_1) = y_1$. We also refer to $y(x)$ as a curve joining the endpoints. Because many curves $y(x)$ join the given endpoints, we add some extra condition to be satisfied that distinguishes a special $y(x)$ from the set of all such curves. Typically, the condition we consider is to minimize some integral along the path. The general problem in the calculus of variations is then the following.

Definition 4.2.1 (Fixed Endpoint Problem). Find the curve $y = y(x)$ with $y(x_0) = y_0$ and $y(x_1) = y_1$ such that the following integral is minimized:

$$J = \int_{x_0}^{x_1} f(x, y(x), y'(x))\, dx.$$

Here $f(x, y(x), y'(x))$ is a function of x, y and $y' = dy/dx$, all thought of as independent variables.

Remark 4.2.2. In fact, we really can't use the techniques presented below to find minima. We can just find the calculus of variations analogue of critical points, so we often say that an integral is *extremized* instead of minimized. See Remark 4.3.4.

The problem which launched the subject of the calculus of variations is mechanical in nature, but provides a paradigm for the subject.

Example 4.2.3 (The Brachistochrone Problem). Given any point (a, b) in the xy-plane, find the curve $y(x)$ joining (a, b) and the origin so that a bead, under the influence of gravity, sliding along a frictionless wire in the shape $y(x)$ from (a, b) to $(0, 0)$, minimizes the time of travel. Indeed, the word brachistochrone is taken from the Greek words *brachist*, which means shortest, and *chronos*, which means time. The problem of the brachistochrone can be cast into a version of the fixed endpoint problem (see Example 4.4.1):

$$\text{Minimize } T = \frac{1}{\sqrt{2g}} \int_a^0 \frac{\sqrt{1 + y'^2}}{\sqrt{b - y}} \, dx$$

with $y(a) = b$ and $y(0) = 0$.

Of course, in geometry itself we want to minimize things as well. For instance, we have the basic problem of finding shortest paths.

Example 4.2.4. We know intuitively that the shortest distance between two points in the plane is attained by a straight line. To see that this is really true, we must solve the problem of determining the curve $y = y(x)$ of minimum arclength joining two points:

$$\text{Minimize } \int \sqrt{1 + y'^2} \, dx.$$

In this case, $f(x, y, y') = \sqrt{1 + y'^2}$.

Calculus of variations problems may require extra constraints as well.

Example 4.2.5 (The Hanging Chain). The potential energy of a hanging chain is proportional to $\int_{x_0}^{x_1} y\sqrt{1 + y'^2} \, dx$, and the equilibrium position of the chain is assumed when the potential energy is minimized. The problem is *constrained*, however, by the fixed length of the chain $L = \int_{x_0}^{x_1} \sqrt{1 + y'^2} \, dx$. Therefore, the equilibrium shape y (i.e., a *catenary*, from the Latin *catena* for chain) will be attained when the integral

$$J = \int_{x_0}^{x_1} y\sqrt{1 + y'^2} \, dx$$

is extremized subject to the constraint (where L is fixed)

$$L = \int_{x_0}^{x_1} \sqrt{1 + y'^2} \, dx.$$

Of course, the catenary is also the shape assumed by telephone wires and used in structurally strong arches — but it is *not* the shape of the cable supporting a suspension bridge (see [**Opr97**]).

Another example of a problem with an extra constraint is the problem we met in § 1.3.8. In order to phrase it, we extend the formulation of the Fixed Endpoint Problem to include curves in the plane which are described parametrically. Suppose a curve is given by a parametrization $(x(t), y(t))$. Then an integral J as in Definition 4.2.1 may be written

$$J = \int_{t_0}^{t_1} f(t, x(t), y(t), \dot{x}(t), \dot{y}(t)) \, dt,$$

where we use the physics notation $\dot{x} = dx/dt$, $\dot{y} = dy/dt$. This is called the *parametric form* or *two-variable form* of the integral.

Example 4.2.6 (The Isoperimetric Problem). Suppose we have a closed curve of fixed length $L = \int \sqrt{\dot{x}^2 + \dot{y}^2} \, dt$ (given in parametric form) and we wish to enclose the *maximum* amount of area. Green's theorem allows us to express area as a line integral

$$\int \frac{1}{2} \left(-y\dot{x} + x\dot{y} \right) dt,$$

so that the problem becomes

$$\text{Maximize} \quad J = \int \frac{1}{2} \left(-y\dot{x} + x\dot{y} \right) dt$$

subject to the constraint (where L is fixed)

$$L = \int \sqrt{\dot{x}^2 + \dot{y}^2} \, dt.$$

Notice that we have phrased this as a two-variable problem in order to use the Green's theorem formulation of area.

Of course, we are interested in minimal surfaces, so the next two examples are especially important to us.

Example 4.2.7 (Least Area Surfaces of Revolution). We have already seen in Theorem 3.2.5 that a minimal surface of revolution is a catenoid. Since least-area surfaces are minimal, this example offers another approach to the question. A surface of revolution has surface area given by

$$A = 2\pi \int y\sqrt{1 + y'^2} \, dx.$$

Therefore, the problem of finding a minimal surface of revolution becomes a problem of minimizing the integral A. See Example 4.4.7 for a solution, and § 5.5 and § 5.6 for Maple analyses.

Example 4.2.8. Let $z = f(x, y)$ be a function of two variables. Surface area is given by the double integral

$$A = \iint \sqrt{1 + f_x^2 + f_y^2} \, dx \, dy.$$

We shall see that the calculus of variations can be done in several independent variables also, so finding minimal surfaces which are graphs of functions reduces to minimizing the integral A. See Example 4.4.8 for a solution, and § 5.11.6 for Maple's solution.

So, we see that many different types of problems fall under the rubric of the calculus of variations. But how do we solve these problems? The next section provides the first step to finding solutions.

4.3. Necessary Conditions: Euler-Lagrange Equations

When we take the derivative, set it equal to zero and solve in ordinary calculus, we obtain critical points which can be maxima, minima or neither. So it is in the calculus of variations as well. With this in mind, let's see how to approach the problem of finding an extremal for an integral.

Suppose $y(x)$ is a curve which minimizes the integral

$$J = \int_{x_0}^{x_1} f(x, y(x), y'(x)) \, dx$$

and let $\hat{y}(x) = y(x) + \epsilon \eta(x)$ be a variation of y. That is, we think of ϵ as being small and we require that $\eta(x_0) = 0$ and $\eta(x_1) = 0$. This means that the curve $\hat{y}(x)$ still joins x_0 and x_1 as well as being 'close' to y. Note too that $\hat{y}' = y' + \epsilon \eta'$. Here's the point. In ordinary calculus, we usually have a function defined in an open interval, and we look for minima where the derivative vanishes. Here we think of J as our function defined on the space of curves joining the given endpoints, and the open interval is replaced by all the curves which are close to $y(x)$; in particular, all of the $\hat{y}(x)$. Now we need to understand how to differentiate J with respect to this space. But that is precisely why we parametrized the $\hat{y}(x)$ by ϵ. We can differentiate

J with respect to the parameter ϵ. So, with this in mind, we think of the integral J as a function of the parameter ϵ:

$$J(\epsilon) = \int_{x_0}^{x_1} f(x, \widehat{y}, \widehat{y}') \, dx = \int_{x_0}^{x_1} f(x, y + \epsilon\, \eta, y' + \epsilon\, \eta') \, dx.$$

In order to recognize the minimum $y(x)$, just as we do in ordinary calculus, we take the derivative of J with respect to ϵ and note that, since $y(x)$ is a minimum by hypothesis, this derivative is zero for $\epsilon = 0$. Note that we can take the derivative inside the integral because, speaking informally, the integral is taken with respect to x and x is independent of ϵ. A precise version of this is given by

Lemma 4.3.1. *If $f(\epsilon, x)$ and $df(\epsilon, x)/d\epsilon$ are continuous, then*

$$\frac{d}{d\epsilon} \int_a^b f(\epsilon, x) \, dx = \int_a^b \frac{df(\epsilon, x)}{d\epsilon} \, dx.$$

Proof. Let $h(z) = \int_a^b f_z(z, x) \, dx$. We want to show that the ϵ-derivative of the integral in question is $h(\epsilon)$. Using the fact that we can switch the order of integration in a double integral (Fubini's theorem), we get

$$\int_c^\epsilon h(z) \, dz = \int_c^\epsilon \int_a^b f_z(z, x) \, dx \, dz$$

$$= \int_a^b \int_c^\epsilon f_z(z, x) \, dz \, dx$$

$$= \int_a^b f(\epsilon, x) \, dx - \int_a^b f(c, x) \, dx.$$

The Fundamental Theorem of Calculus then gives

$$h(\epsilon) = \frac{d}{d\epsilon} \int_c^\epsilon h(z) \, dz$$

$$= \frac{d}{d\epsilon} \int_a^b f(\epsilon, x) \, dx - \frac{d}{d\epsilon} \int_a^b f(c, x) \, dx$$

$$= \frac{d}{d\epsilon} \int_a^b f(\epsilon, x) \, dx$$

since the second term does not depend on ϵ. This is of course exactly what we wanted. □

Now, we have

$$\frac{dJ}{d\epsilon} = \int_{x_0}^{x_1} \frac{\partial f}{\partial \widehat{y}} \frac{\partial \widehat{y}}{\partial \epsilon} + \frac{\partial f}{\partial \widehat{y}'} \frac{\partial \widehat{y}'}{\partial \epsilon} \, dx$$

by the chain rule (since x does not depend on ϵ), so

$$\frac{dJ}{d\epsilon} = \int_{x_0}^{x_1} \frac{\partial f}{\partial \widehat{y}} \eta + \frac{\partial f}{\partial \widehat{y}'} \eta' \, dx$$

with

$$0 = \left. \frac{dJ}{d\epsilon} \right|_{\epsilon=0} = \int_{x_0}^{x_1} \frac{\partial f}{\partial y} \eta + \frac{\partial f}{\partial y'} \eta' \, dx$$

since, at $\epsilon = 0$, $\widehat{y}(x) = y(x)$. Now, the second term inside the integral may be integrated by parts as follows. Let

$$u = \frac{\partial f}{\partial y'}, \qquad\qquad dv = \eta' \, dx,$$

$$du = \frac{d}{dx}\left(\frac{\partial f}{\partial y'}\right) dx, \qquad\qquad v = \eta,$$

and compute

$$\int_{x_0}^{x_1} \frac{\partial f}{\partial y'} \eta' \, dx = \left.\eta \frac{\partial f}{\partial y'}\right|_{x_0}^{x_1} - \int_{x_0}^{x_1} \eta \frac{d}{dx}\left(\frac{\partial f}{\partial y'}\right) dx$$

$$= -\int_{x_0}^{x_1} \eta \frac{d}{dx}\left(\frac{\partial f}{\partial y'}\right) dx$$

since $\eta(x_0) = 0 = \eta(x_1)$ implies the vanishing of the first term. Putting this in the integral above, we have

$$0 = \int_{x_0}^{x_1} \frac{\partial f}{\partial y} \eta \, dx - \int_{x_0}^{x_1} \eta \frac{d}{dx}\left(\frac{\partial f}{\partial y'}\right) dx$$

$$= \int_{x_0}^{x_1} \eta \left[\frac{\partial f}{\partial y} - \frac{d}{dx}\left(\frac{\partial f}{\partial y'}\right)\right] dx.$$

This equation must hold for every function η with $\eta(x_0) = 0 = \eta(x_1)$.

Exercise 4.3.2. Suppose that a continuous function $y = f(x)$ is positive at a point x_0. Show that it is possible to choose a function $\eta(x)$ so that on some interval $[a, b]$

$$\int_a^b \eta(x) f(x) \, dx > 0.$$

Hints: (1) draw a graph of $f(x)$ about x_0 and use continuity to say that there is a whole interval of x's around x_0 such that $f(x) > 0$; (2) create a 'bump' function $\eta(x)$ which is non-zero only inside the interval of x's in part (1) – this guarantees the positivity of the integrand in the integral above; (3) now remember that positive integrands produce positive integrals.

The exercise tells us that there is only one way the equation

$$0 = \int_{x_0}^{x_1} \eta \left[\frac{\partial f}{\partial y} - \frac{d}{dx} \left(\frac{\partial f}{\partial y'} \right) \right] dx$$

can hold for all choices of η. We must have

$$\frac{\partial f}{\partial y} - \frac{d}{dx} \left(\frac{\partial f}{\partial y'} \right) = 0.$$

This equation is called the *Euler-Lagrange equation* and it gives us a *necessary condition* for $y(x)$ to be a minimizer for J, just as the vanishing of the ordinary first derivative does for a function in 1-variable calculus.

Theorem 4.3.3. *If $y = y(x)$ is a minimizer for the fixed endpoint problem, then y satisfies the Euler-Lagrange equation.*

Remark 4.3.4. Notice that we are *not* saying that a solution to the Euler-Lagrange equation is a solution to the fixed endpoint problem. The Euler-Lagrange equation is simply a first step toward solving the fixed endpoint problem, akin to finding critical points in calculus. In fact, however, sometimes the *only* solutions which can be readily found are solutions to the Euler-Lagrange equations. For this reason, solutions to the Euler-Lagrange equation are given the special name *extremals* and the fixed endpoint problem, for example, is often rephrased to say that a curve $y(x)$ is desired which joins the given endpoints and *extremizes* the integral J. In this case, solutions to the Euler-Lagrange equation solve the problem.

Example 4.3.5. Suppose that we have a problem in parametric form. That is, J depends on two functions $x(t)$ and $y(t)$ which each depend on an independent variable t:

$$J = \int_{t_0}^{t_1} f(t, x(t), y(t), \dot{x}(t), \dot{y}(t)) \, dt.$$

We can then take two variations $\hat{x} = x + \epsilon\eta$ and $\hat{y} = y + \epsilon\tau$ which lead to

$$
\begin{aligned}
0 &= \frac{dJ}{d\epsilon}\bigg|_{\epsilon=0} \\
&= \int_{t_0}^{t_1} \frac{\partial f}{\partial x}\frac{\partial x}{\partial \epsilon} + \frac{\partial f}{\partial \dot{x}}\frac{\partial \dot{x}}{\partial \epsilon} + \frac{\partial f}{\partial y}\frac{\partial y}{\partial \epsilon} + \frac{\partial f}{\partial \dot{y}}\frac{\partial \dot{y}}{\partial \epsilon} \, dt \\
&= \int_{t_0}^{t_1} \frac{\partial f}{\partial x}\eta + \frac{\partial f}{\partial \dot{x}}\dot{\eta} + \frac{\partial f}{\partial y}\tau + \frac{\partial f}{\partial \dot{y}}\dot{\tau} \, dt \\
&= \int_{t_0}^{t_1} \eta\left[\frac{\partial f}{\partial x} - \frac{d}{dt}\left(\frac{\partial f}{\partial \dot{x}}\right)\right] dt + \int_{t_0}^{t_1} \tau\left[\frac{\partial f}{\partial y} - \frac{d}{dt}\left(\frac{\partial f}{\partial \dot{y}}\right)\right] dt.
\end{aligned}
$$

This equation holds for all η, τ and, since we can let $\eta = 0$ or $\tau = 0$ independently, the argument for the single function case says that the following *Euler-Lagrange equations* hold:

$$\frac{\partial f}{\partial x} - \frac{d}{dt}\left(\frac{\partial f}{\partial \dot{x}}\right) = 0, \qquad \frac{\partial f}{\partial y} - \frac{d}{dt}\left(\frac{\partial f}{\partial \dot{y}}\right) = 0.$$

Exercise 4.3.6. Suppose that J depends on two independent variables t and s,

$$J = \iint_{\mathcal{R}} f(t, s, x(t, s), x_t(t, s), x_s(t, s)) \, dt \, ds.$$

Here we think of $x(t, s)$ either as a function of two variables or as a parametrization of a surface. Show that a variation $\hat{x}(t, s) = x(t, s) + \epsilon\eta(t, s)$ with $\eta|_C = 0$, where C is the boundary of the region of integration $\mathcal{R}$, leads to

$$0 = \frac{dJ}{d\epsilon}\bigg|_{\epsilon=0} = \iint \frac{\partial f}{\partial x}\eta + \frac{\partial f}{\partial x_t}\eta_t + \frac{\partial f}{\partial x_s}\eta_s \, dt \, ds.$$

Further, recalling Green's theorem

$$\int -P\,dt + Q\,ds = \iint \frac{\partial Q}{\partial t} + \frac{\partial P}{\partial s}\,dt\,ds,$$

letting $Q = \eta\,(\partial f/\partial x_t)$, $P = \eta\,(\partial f/\partial x_s)$ and using $\eta|_C = 0$, show that the last two terms of the integral give

$$\iint \frac{\partial f}{\partial x_t}\eta_t + \frac{\partial f}{\partial x_s}\eta_s\,dt\,ds = -\iint \eta\left[\frac{\partial^2 f}{\partial x_t \partial t} + \frac{\partial^2 f}{\partial x_s \partial s}\right]\,dt\,ds,$$

and substitution then gives

$$0 = \iint_{\mathcal{R}} \eta\left[\frac{\partial f}{\partial x} - \frac{\partial^2 f}{\partial x_t \partial t} - \frac{\partial^2 f}{\partial x_s \partial s}\right]\,dt\,ds.$$

Finally, argue that this implies that the *Euler-Lagrange equation* for two independent variables is

$$\frac{\partial f}{\partial x} - \frac{\partial}{\partial t}\left(\frac{\partial f}{\partial x_t}\right) - \frac{\partial}{\partial s}\left(\frac{\partial f}{\partial x_s}\right) = 0.$$

Example 4.3.7 (A Special Case of Plateau's 90° Rule). We have already seen in Example 3.9.13 how Plateau's 90° rule may be proved via complex variables. Let's see what the calculus of variations has to say. In Exercise 4.3.6 Green's theorem produces a line integral

$$\int_C \eta\,\frac{\partial f}{\partial x_t}\,ds - \eta\,\frac{\partial f}{\partial x_s}\,dt,$$

which vanishes because $\eta|_C = 0$. But what if the curve C consists of two parts: one, C_1, where $\eta|_{C_1} = 0$ and one, C_2, where $\eta|_{C_2} \neq 0$, but where the surface given by the Monge parametrization $(t, s, x(t, s))$ (restricted to C_2) is required to be on a plane. This is the free boundary in a plane mentioned in Plateau's 90° rule (see § 1.5). With this assumption, the derivation outlined in Exercise 4.3.6 must be modified. Namely, instead of

$$0 = \iint_{\mathcal{R}} \eta\left[\frac{\partial f}{\partial x} - \frac{\partial^2 f}{\partial x_t \partial t} - \frac{\partial^2 f}{\partial x_s \partial s}\right]\,dt\,ds,$$

we have

$$\int_{C_2} \eta \frac{\partial f}{\partial x_t} \, ds - \eta \frac{\partial f}{\partial x_s} \, dt - \iint_{\mathcal{R}} \eta \left[\frac{\partial f}{\partial x} - \frac{\partial^2 f}{\partial x_t \partial t} - \frac{\partial^2 f}{\partial x_s \partial s} \right] dt \, ds$$

$$= \iint \frac{\partial f}{\partial x_t} \eta_t + \frac{\partial f}{\partial x_s} \eta_s \, dt \, ds.$$

After substituting in the usual integral, we obtain

$$\iint \eta \left[\frac{\partial f}{\partial x} - \frac{\partial}{\partial t} \left(\frac{\partial f}{\partial x_t} \right) - \frac{\partial}{\partial s} \left(\frac{\partial f}{\partial x_s} \right) \right] ds \, dt$$

$$+ \int_{C_2} \eta \frac{\partial f}{\partial x_t} \, ds - \eta \frac{\partial f}{\partial x_s} \, dt = 0.$$

Now, once the final extremal is obtained with a fixed boundary curve on the plane, it is an extremal for the fixed boundary problem (i.e. where $\eta = 0$ on the entire boundary). Hence, the Euler-Lagrange equation of Exercise 4.3.6 must be satisfied, and this means that the double integral above vanishes. So, we are left with

$$\int_{C_2} \eta \frac{\partial f}{\partial x_t} \, ds - \eta \frac{\partial f}{\partial x_s} \, dt = 0.$$

As usual, since this equation holds for all η, we really are saying that, on the curve C_2 (parametrized by r with $\dot{s} = ds/dr$ and $\dot{t} = dt/dr$),

$$\frac{\partial f}{\partial x_t} \dot{s} - \frac{\partial f}{\partial x_s} \dot{t} = 0.$$

Now we specialize to the case where f represents area;

$$f = \sqrt{1 + x_t^2 + x_s^2}.$$

Then we have

$$\frac{\partial f}{\partial x_s} = \frac{x_s}{\sqrt{1 + x_s^2 + x_t^2}}, \qquad \frac{\partial f}{\partial x_t} = \frac{x_t}{\sqrt{1 + x_s^2 + x_t^2}},$$

and substituting in the equation above gives

$$\frac{x_t}{\sqrt{1 + x_t^2 + x_s^2}} \dot{s} - \frac{x_s}{\sqrt{1 + x_t^2 + x_s^2}} \dot{t} = 0.$$

Only the numerator matters, so we have $x_t \dot{s} - x_s \dot{t} = 0$.

Suppose that the plane is vertical for convenience. If not, the whole situation may be rotated to achieve this without changing the geometry (but possibly changing $\mathbf{x}(u,v)$ near the plane). The normal for the plane is then given by $n = (a, b, 0)$, while the calculations at the start of § 3.2 show that the Monge parametrization $\mathbf{x} = (t, s, x(t, s))$ has

$$\mathbf{x}_t = (1, 0, x_t), \quad \mathbf{x}_s = (0, 1, x_s), \quad U = \frac{(-x_t, -x_s, 1)}{\sqrt{1 + x_t^2 + x_s^2}}.$$

Then, if we denote the intersection curve of the surface $\mathbf{x}$ with the plane by α, we have by Lemma 2.1.2 that $\dot{\alpha} = \dot{t}\mathbf{x}_t + \dot{s}\mathbf{x}_s = \dot{u}(1, 0, x_t) + \dot{v}(0, 1, x_s)$, where we still use the 'dot' notation for differentiation with respect to the curve's parameter. Because α is in the plane, its tangent vector is perpendicular to n: $0 = a\dot{t} + b\dot{s}$. Also, we have $U \cdot n = -ax_t - bx_s$. Substituting $b = -a\dot{t}/\dot{s}$ into this equation gives

$$U \cdot n = -ax_t - \left(\frac{-a\dot{t}}{\dot{s}}\right) x_s = -\frac{a}{\dot{s}}(\dot{s}x_t - \dot{t}x_s) = 0$$

since $x_s \dot{t} - x_t \dot{s} = 0$. But then the minimal surface is perpendicular to the plane, as predicted by Plateau.

Exercise 4.3.8. Suppose $f(x, y, y') = f(y, y')$ does not depend on x explicitly. By this, we mean that f only depends on x through the curves y and y', so that $\partial f / \partial x = 0$. Show that a non-constant $y(x)$ satisfies the Euler-Lagrange equation if and only if

$$f - y' \frac{\partial f}{\partial y'} = c,$$

where c is a constant. Hint: compute the derivative with respect to x of the left-hand side. Don't forget the chain rule. This equation is sometimes called a *first integral* for f.

Exercise 4.3.9. Suppose $f(t, x, y, \dot{x}, \dot{y}) = f(x, y, \dot{x}, \dot{y})$ does not depend on t explicitly. By this, we mean that f only depends on t through the curves x, y, $\dot{x}$ and $\dot{y}$, so that $\partial f / \partial t = 0$. Show that non-constant $x(t)$, $y(t)$ satisfy the Euler-Lagrange equations if and only if

$$f - \dot{x} \frac{\partial f}{\partial \dot{x}} - \dot{y} \frac{\partial f}{\partial \dot{y}} = c,$$

where c is a constant. Hint: compute the derivative with respect to t of the left-hand side. Don't forget the chain rule.

Exercise 4.3.10. Show that, for $f = f(x, y, y')$, if $\partial f/\partial y = 0$, then $y(x)$ satisfies the Euler-Lagrange equation if and only if $\partial f/\partial y' = c$, a constant.

Example 4.3.11. Find the extremal $y(x)$ for the fixed endpoint problem

$$J = \int_0^\pi y \sin x + \frac{1}{2} {y'}^2 \, dx$$

with $y(0) = 0$, $y(\pi) = \pi$. The Euler-Lagrange equation is

$$\sin(x) - \frac{d}{dx}(y') = 0.$$

This gives $y'' = \sin(x)$ and integrating twice produces

$$y(x) = -\sin(x) + cx + d.$$

The boundary conditions can now be used to determine c and d. The condition $y(0) = 0$ immediately gives $d = 0$, and the condition $y(\pi) = \pi$ then gives $c = 1$. Hence, the extremal is $y(x) = -\sin(x) + x$. After doing the two problems below, see § 5.11.

Exercise 4.3.12. Find the extremal $y(x)$ for the fixed endpoint problem

$$J = \int_0^{\pi/2} {y'}^2 - y^2 \, dx$$

with $y(0) = 0$, $y(\pi/2) = 1$. Do this using the ordinary Euler-Lagrange equation and using the first integral. You should get $y(x) = \sin(x)$. See § 5.11 also.

Exercise 4.3.13. Find the extremal $y(x)$ for the fixed endpoint problem

$$J = \int_0^1 y' y^2 + {y'}^2 y \, dx$$

with $y(0) = 1$, $y(1) = 4$. Do this first by the first integral and then by the Euler-Lagrange equation. Finally, see § 5.11.

Exercise 4.3.14. Here is a type of problem where it is possible to identify an extremal as a minimizer (as well as calculate the value of J in a simple way). Consider the fixed endpoint problem with $y(x_0) = y_0$, $y(x_1) = y_1$ and integral to be minimized

$$J = \int_{x_0}^{x_1} p^2 y'^2 + q^2 y^2 \, dx,$$

where $p = p(x)$ and $q = q(x)$ are arbitrary smooth functions.

(1) Show that if $y = y(x)$ is an extremal for J, then

$$J = p^2 y y'|_{x_0}^{x_1} = p^2 y y'(x_1) - p^2 y y'(x_0).$$

Thus the value of J is easy to calculate for an extremal.

(2) Show that if $y = y(x)$ is an extremal for J and $\eta = \eta(x)$ is a function with $\eta(x_0) = 0 = \eta(x_1)$, then

$$\widetilde{J} = \int_{x_0}^{x_1} p^2 y' \eta' + q^2 y \eta \, dx = 0.$$

(3) From (1) and (2) show that an extremal $y(x)$ for J is a solution to the fixed endpoint problem. That is, $y(x)$ minimizes J.

Hints: For (1), differentiate $y p^2 y'$ with respect to x using the product rule on $(y)(p^2 y')$ as indicated, and remember that y is an extremal. For (2), compute $\int p^2 y' \eta' \, dx$ by parts, remembering that y is an extremal, or review the derivation of the Euler-Lagrange equation. For (3), vary y by $y + \eta$ (where the usual ϵ is absorbed into η) and expand the integrand of

$$\widehat{J} = \int_{x_0}^{x_1} p^2 (y' + \eta')^2 + q^2 (y + \eta)^2 \, dx.$$

Note $\widehat{J} = J + \dots$. What can you say about the ...?

4.4. Solving the Fundamental Examples

Now let's actually solve some of the problems we have discussed using the Euler-Lagrange equation. Because the problem of determining the *brachistochrone* is the oldest problem in the subject, it deserves to be treated first. Recall that the problem itself is this:

Example 4.4.1 (The Brachistochrone Problem). Given any point (a, b) in the xy-plane, find the curve $y(x)$ joining (a, b) and the origin so that a bead, under the influence of gravity, sliding along a frictionless wire in the shape $y(x)$ from (a, b) to $(0, 0)$, minimizes the time of travel. The key to setting up this problem is the simple formula $D = RT$, distance equals rate times time. The 'infinitesimal' distance travelled by the bead is just the arclength of the wire, $D = \sqrt{1 + y'^2}$. The rate of descent may be determined by noting that potential energy is given by the formula mgh, where m is the mass of the bead, $g = 9.8\,\text{m/sec}^2$ is the acceleration due to gravity (near the surface of the Earth) and h is the height of the bead above a fixed reference height (which is often taken to be zero on the Earth's surface). If we start the bead at (a, b) with initial velocity zero, then conservation of energy requires that the kinetic energy of the bead be equal to the loss of potential energy due to decreasing height. In other words,

$$mg(b - y) = \frac{1}{2}mv^2,$$

where v is the speed of the bead. We then have $v = \sqrt{2g(b - y)}$. So, $D = RT$ tells us that the infinitesimal time dT is a function of y and y',

$$dT(y, y') = \frac{1}{\sqrt{2g}} \frac{\sqrt{1 + y'^2}}{\sqrt{b - y}}.$$

The total time is then found by integrating $dT(y, y')$ with respect to x from a to 0. The problem of the brachistochrone now becomes the fixed endpoint problem

$$\text{Minimize } T = \frac{1}{\sqrt{2g}} \int_a^0 \frac{\sqrt{1 + y'^2}}{\sqrt{b - y}}\, dx$$

with $y(a) = b$ and $y(0) = 0$.

To solve this problem, first substitute $u = b - y$ and ignore the constant factor $1/\sqrt{2g}$ (since it can't make a difference in whether the derivative of a quantity is zero or not) to get a new integral to be extremized,

$$J = \int_a^0 \frac{\sqrt{1 + u'^2}}{\sqrt{u}}\, dx.$$

Because the independent variable does not appear explicitly in J, we use the first integral (Exercise 4.3.8) to get

$$\frac{\sqrt{1+u'^2}}{\sqrt{u}} - u' \frac{u'}{\sqrt{u(1+u'^2)}} = c.$$

Now, finding a common denominator, simplifying and replacing $1/c$ by a new constant $\sqrt{c}$ gives a separable differential equation $c = u(1+u'^2)$, which leads to

$$x = \int \frac{\sqrt{u}}{\sqrt{c-u}}\, du$$

$$= -2 \int \sqrt{c^2 - w^2}\, dw,$$

where $w = \sqrt{c-u}$ and c is replaced by c^2,

$$= -2 \int c\cos(\theta)\, c\cos(\theta)\, d\theta,$$

where $w = c\sin(\theta)$,

$$= -2c^2 \int \cos^2(\theta)\, d\theta$$

$$= -2c^2 \int \frac{1+\cos(2\theta)}{2}\, d\theta$$

$$= -\frac{c^2}{2}[2\theta + \sin(2\theta)] + r,$$

and the final answer

$$x(\phi) = k(\phi + \sin(\phi)) + r,$$

where $2\theta = \phi$. Now, $b - y = u = c^2\cos^2(\theta) = c^2\cos^2(\phi/2)$ by the substitutions above, so, again using $\cos^2(\phi/2) = (1+\cos(\phi))/2$, we obtain

$$y(\phi) = k(1 + \cos(\phi)) + b,$$

where $k = -c^2/2$ as for $x(\phi)$. Note that $y(\pi) = b$, so that it must also be true that $x(\pi) = a$. This implies that $r = a - k\pi$. If we let $\zeta = \phi - \pi$, then we obtain the parametric formulas for x and y,

$$x(\zeta) = k(\zeta - \sin(\zeta)) + a \quad \text{and} \quad y(\zeta) = k(1 - \cos(\zeta)) + b.$$

This is the parametrization of a cycloid. Hence, the solution to the problem of the brachistochrone is a standard curve. For a Maple race between a bead on a line and one on a cycloid, see § 5.12. For the *brachistochrone with friction*, see [**HK95**], and for a very different approach, see [**Law96**].

Example 4.4.2 (Bernoulli's Brachistochrone Solution). Johann Bernoulli solved the brachistochrone problem ingeniously by employing Fermat's principle that light travels to minimize time together with Snell's law of refraction (see [**Wei74**] and Example 4.1.1). The idea is this. Consider the path of the curve as passing through contiguous (horizontal) layers of media with different refraction indices (meaning that light has different speeds in the layers). We think of the path of the bead as the path a photon would follow through the media. Also, approximate the photon path within each layer by a straight line. The same energy arguments as in Example 4.4.1 show that the speed in layer i is given by $\sqrt{2gy_i}$ (where we assume the speed in each layer is constant, as it is for light of course). If we let α_i be the angle between the curve and the normal (which is vertical here) in layer i, then Snell's law says that

$$\frac{\sin(\alpha_1)}{\sqrt{y_1}} = \frac{\sin(\alpha_2)}{\sqrt{y_2}} = \ldots = \frac{\sin(\alpha_n)}{\sqrt{y_n}} = c,$$

a constant. Taking the limit as the layers get thinner, we obtain an equation $\sin(\alpha(x))/\sqrt{y(x)} = c$.

Now, we know that the slope of a line is the tangent of the angle made with the horizontal axis, so the tangent line to the brachistochrone has slope

$$y' = \cotan(\alpha) = \frac{1}{\tan(\alpha)}$$

since α is the angle made with the vertical axis. We then have

$$\sec^2(\alpha) = 1 + \tan^2(\alpha)$$
$$= 1 + \frac{1}{y'^2},$$
$$\cos^2(\alpha) = \frac{y'^2}{1 + y'^2}.$$

Then we easily get

$$\sin(\alpha) = \sqrt{1 - \cos^2(\alpha)}$$
$$= \frac{1}{\sqrt{1 + y'^2}}.$$

Combining this with $\sin(\alpha) = c\sqrt{y}$, we get the differential equation

$$y(1 + y'^2) = c,$$

and this was the equation derived from the Euler-Lagrange equation in Example 4.4.1. Hence, the final answer is the same — the cycloid. This whole approach to mechanical problems is known as the *optical mechanical analogy*.

Exercise 4.4.3. Consider a cycloid of the form

$$(x(t), y(t)) = (A + a(t - \sin t), \ B - a(1 - \cos t)).$$

Graph this cycloid and compare it to the one found in Exercise 4.4.1. Suppose a unit mass particle starts at rest at a point $(\bar{x}, \bar{y})$ on the cycloid corresponding to an angle $\bar{t}$ in the parametrization above. Under the influence of gravity (g being the gravitational constant, and assuming no friction), no matter what initial $\bar{t}$ is chosen, it always takes a time of

$$T = \sqrt{\frac{a}{g}} \pi$$

for the particle to slide down to the bottom of the cycloid (i.e. $t = \pi$). This property was used by C. Huygens to make clocks with pendula extensible along a cycloid, therefore ensuring a constant swing time (which he hoped would allow for accurate timekeeping on ships at sea, thereby allowing the precise calculation of longitude). He named the cycloid the *tautochrone* since this word means 'same time'. Hints: potential energy is turned into kinetic energy by $v^2/2 = g(\bar{y} - y)$.

Time equals distance divided by speed, so

$$T = \frac{1}{\sqrt{2g}} \int_{\bar{x}}^{x_{\text{bot}}} \frac{\sqrt{1 + y'^2}}{\sqrt{\bar{y} - y}} \, dx$$

$$= \sqrt{\frac{a}{g}} \int_{\bar{t}}^{\pi} \sqrt{\frac{1 - \cos t}{\cos \bar{t} - \cos t}} \, dt$$

$$= \sqrt{\frac{a}{g}} \Big|_{\bar{t}}^{\pi} - 2 \arcsin \left[\frac{\sqrt{2} \cos \frac{t}{2}}{\sqrt{1 + \cos \bar{t}}} \right]$$

using

$$y' = \frac{(dy/dt)}{(dx/dt)} = \frac{-\sin t}{(1 - \cos t)},$$

$$\bar{y} = B - a(1 - \cos \bar{t}),$$

$$y = B - a(1 - \cos t),$$

and simplifying. Verify the final step by differentiation, and compute T.

Exercise 4.4.4. Here's a problem in the same vein as Bernoulli's approach to the brachistochrone Example 4.4.2. Suppose that a light photon moves in the upper half-plane to minimize time. Suppose also that the plane is made of a medium whose properties imply that the speed of the photon is always proportional to its height above the x-axis. That is, $v = k\, y$ for $k > 0$. What path does the photon take? Hint: set up the time integral and extremize it. This problem may be related to the paths seismic waves take under the surface of the Earth [Tro96] as well as to the reason we see mirages in deserts [Lem97]. Finally, the solutions to this problem are geodesics in the Poincaré plane. Does this have any physical meaning?

Example 4.4.5 (Shortest distance curves in the plane). For given points (a, b) and (c, d) in the plane, we can ask for the curve $(x(t), y(t))$ which joins them and which minimizes the arclength integral

$$J = \int_{t_0}^{t_1} \sqrt{\dot{x}^2(t) + \dot{y}^2(t)} \, dt.$$

Here we have chosen a two-variable formulation to show how these types of problems work. Because the integrand does not depend on either x or y, the Euler-Lagrange equations reduce to

$$\frac{\partial f}{\partial \dot{x}} = \frac{\dot{x}}{\sqrt{\dot{x}^2(t) + \dot{y}^2(t)}} = c \quad \text{and} \quad \frac{\partial f}{\partial \dot{y}} = \frac{\dot{y}}{\sqrt{\dot{x}^2(t) + \dot{y}^2(t)}} = d.$$

Solving for the square root in each equation and setting the results equal, we obtain

$$\frac{\dot{y}}{d} = \frac{\dot{x}}{c},$$

which gives, under integration with respect to t,

$$y = \frac{d}{c}x + r.$$

This is, of course, the equation of a straight line in the plane. Therefore, extremals of the arclength integral are straight lines.

Exercise 4.4.6. Solve the shortest distance in the plane problem using one variable. That is, arclength is given by

$$\int \sqrt{1 + y'(x)^2}\, dx.$$

Example 4.4.7 (Least area surfaces of revolution). Recall from calculus that a surface of revolution has surface area given by

$$A = 2\pi \int y\sqrt{1 + y'^2}\, dx.$$

Since the integrand of this surface area integral is independent of x, we can apply Exercise 4.3.8 to get the following differential equation for an extremal:

$$y\sqrt{1 + y'^2} - y'\left(\frac{yy'}{\sqrt{1 + y'^2}}\right) = c.$$

This is easily solved for y' to get $y' = \sqrt{y^2 - c^2}/c$, a separable differential equation whose solution is $y = c\cosh(x/c - d)$, a catenary. This agrees with the fact that a catenoid is the only minimal surface of revolution. See § 5.11.4 for a Maple approach to these calculations. We have already seen that a catenoid is least area for a given boundary only under certain circumstances, so this example shows again that extremals are not always minimizers.

Example 4.4.8. Let $z = f(x, y)$ be a function of two variables. By the calculations in Example 3.3.3, we know that the surface area of the graph of f is given by

$$\text{Area} = \iint \sqrt{1 + f_u^2 + f_v^2} \, du \, dv,$$

where $\mathbf{x}(u, v) = (u, v, f(u, v))$ is a Monge parametrization. Extremizing surface area is then a two-variable calculus of variations problem, and the Euler-Lagrange equation from Exercise 4.3.6 applies. We get

$$0 = \frac{\partial(\sqrt{1 + f_u^2 + f_v^2})}{\partial f}$$

$$- \frac{\partial}{\partial u}\left(\frac{\partial(\sqrt{1 + f_u^2 + f_v^2})}{\partial f_u}\right)$$

$$- \frac{\partial}{\partial v}\left(\frac{\partial(\sqrt{1 + f_u^2 + f_v^2})}{\partial f_v}\right)$$

$$= 0 - \frac{\partial}{\partial u}\left(\frac{f_u}{\sqrt{1 + f_u^2 + f_v^2}}\right) - \frac{\partial}{\partial v}\left(\frac{f_v}{\sqrt{1 + f_u^2 + f_v^2}}\right)$$

$$= \frac{f_{uu}(1 + f_u^2 + f_v^2) - f_u(f_u f_{uu} + f_v f_{uv})}{(1 + f_u^2 + f_v^2)^{3/2}}$$

$$+ \frac{f_{vv}(1 + f_u^2 + f_v^2) - f_v(f_u f_{uv} + f_v f_{vv})}{(1 + f_u^2 + f_v^2)^{3/2}}$$

$$= \frac{f_{uu}(1 + f_v^2) - 2f_u f_v f_{uv} + f_{vv}(1 + f_u^2)}{(1 + f_u^2 + f_v^2)^{3/2}}.$$

Taking the numerator of the fraction, we obtain

$$f_{uu}(1 + f_v^2) - 2f_u f_v f_{uv} + f_{vv}(1 + f_u^2) = 0.$$

This is, of course, the minimal surface equation, so we have another proof of Theorem 3.3.6 using the Euler-Lagrange equation instead of direct analysis. Also see Example 3.10.1 for situations where true minimization of surface area can be proved, and § 5.11.6 for Maple's derivation.

These last two examples are the heart and soul of the calculus of variations applied to minimal surfaces and soap films. Notice that, while the general function-of-two-variables case gives a complicated partial differential equation in two variables, the rotational symmetry of the surface of revolution case reduces the problem to an ordinary, and even separable, differential equation. This may be the first case in history of symmetry reducing the order of a physical problem. Now let's return for a moment to harmonic functions and see how they arise from the technique of the calculus of variations.

Exercise 4.4.9 ([**CH53**]). Suppose $z = f(x, y)$ is a function whose graph spans a given curve and has least area among surfaces spanning the curve. Further, suppose all higher powers ≥ 4 of partials of f are negligible. That is, f_x^4, f_y^4 and $f_x^2 f_y^2$ are so small that the integrand $\sqrt{1 + f_x^2 + f_y^2}$ of the area integral is well approximated by $\sqrt{(1 + \frac{1}{2}[f_x^2 + f_y^2])^2} = 1 + \frac{1}{2}[f_x^2 + f_y^2]$. Then, up to a constant provided by the term 1 and the factor $1/2$, the relevant area integral is

$$A = \int f_x^2 + f_y^2 \, dx \, dy.$$

Now vary $z = f(x, y)$ to get a nearby surface $z^t(x, y) = f(x, y) + tg(x, y)$, and carry through the derivation of Theorem 3.3.6 as before using the area integral above to show that f is harmonic. Unfortunately, in most cases, as the minimal surface becomes less planar (i.e. when fourth-order terms are no longer negligible), it begins to vary from the graph of a true harmonic function.

Exercise 4.4.10. This exercise is a companion to Exercise 4.4.9, since it uses the Euler-Lagrange equation to find extremals for the integral presented there. Let $z = f(x, y)$ be a function of two variables. The *Dirichlet integral* is given by

$$A = \iint f_x^2 + f_y^2 \, dx \, dy.$$

Use the two-independent-variable Euler-Lagrange equation of Exercise 4.3.6 to show that an extremal for this integral is a harmonic function. Dirichlet's integral arises in many areas of physics and engineering. In particular, the two-variable version above could represent the capacity (per unit length) of a cylindrical condenser with annular cross-sections, and the analogous three-variable version could represent the potential energy associated with an electric field. For the latter, a stable equilibrium is attained when potential energy is minimized, so a minimizer for the Dirichlet integral is identified with the potential for the field. For more information, see [**Wei74**].

Once again, let's note that extremals are most certainly *only possible* minimizers. A true analysis of sufficiency conditions for minimizers is too complicated to be presented here, but we sampled a bit of the flavor of the subject in § 3.10. Here is a simple example of something that can go wrong.

Exercise 4.4.11. For the fixed endpoint problem with $y(0) = 1$, $y(1) = 0$ and

$$J[y] = \int_0^1 y^4 + y\,y' + \frac{1}{2} \; dx,$$

a) Show that the only extremal is $y(x) = 0$, so that there is *no* minimizer (since the endpoint conditions are not satisfied).
b) Show that $J[y] \geq 0$. Hint: show that $J[y] = \int_0^1 y^4 \, dx$. Use $y'\, dx = dy$.
c) Show that $J[y]$ gets arbitrarily close to 0. Hint: consider the piecewise linear function defined by

$$y(x) = \begin{cases} 1 - \epsilon\, x, & 0 \leq x \leq \frac{1}{\epsilon}, \\ 0, & \frac{1}{\epsilon} \leq x \leq 1, \end{cases}$$

and compute $J[y]$ for these functions. Therefore, the minimum for $J[y]$ cannot be any number greater than zero. But the only y which gives $J[y] = 0$ is $y = 0$, and this cannot be the solution by a) above. Thus, this problem has *no* solution. This illustrates a common problem in the calculus of variations. The Euler-Lagrange equation is a necessary condition for a solution, but it does not guarantee the existence of a solution.

4.5. Problems with Extra Constraints

Now let's look at those variational problems having extra constraints, like the hanging chain and the isoperimetric problem. Let's set up a standard problem for this situation. It shall be:

$$\text{Minimize } J = \int_{x_0}^{x_1} f(x, y, y') \, dx$$

subject to the endpoint conditions $y(x_0) = y_0$, $y(x_1) = y_1$ and the requirement that

$$I = \int_{x_0}^{x_1} g(x, y, y') \, dx = c,$$

where c is a constant.

We will now derive a form of the Euler-Lagrange equation which is a necessary condition for the solution of the constrained problem. Just as before, assume $y = y(x)$ is a minimizer for the problem and take a variation $\widehat{y} = y + \epsilon (a\eta + b\xi)$, where $\eta(x_0) = 0 = \eta(x_1)$ and $\xi(x_0) = 0 = \xi(x_1)$. We take two 'perturbations' η and ξ, because taking only one would not allow us to vary J while holding I constant. By taking η *and* ξ, we can vary J while offsetting the effects of one perturbation with the other in I. The usual Euler-Lagrange argument gives

$$0 = \frac{dJ}{d\epsilon}\bigg|_{\epsilon=0} = \int_{x_0}^{x_1} (a\eta + b\xi) \left(\frac{\partial f}{\partial y} - \frac{d}{dx} \left(\frac{\partial f}{\partial y'} \right) \right) dx.$$

Now, however, the usual argument must be modified because $a\eta + b\xi$ is *not* arbitrary. The requirement $I = c$ puts restrictions on a and b. If we carry through the argument above for I, we get

$$0 = \frac{dI}{d\epsilon}\bigg|_{\epsilon=0} = \int_{x_0}^{x_1} (a\eta + b\xi) \left(\frac{\partial g}{\partial y} - \frac{d}{dx} \left(\frac{\partial g}{\partial y'} \right) \right) dx$$

where the derivative is zero since $I = c$ is a constant. But y is *not* an extremal for I, so the Euler-Lagrange equation with g does not hold. Instead, we have

$$\int_{x_0}^{x_1} (a\eta + b\xi) \left(\frac{\partial f}{\partial y} - \frac{d}{dx} \left(\frac{\partial f}{\partial y'} \right) \right) dx = 0,$$

$$\int_{x_0}^{x_1} (a\eta + b\xi) \left(\frac{\partial g}{\partial y} - \frac{d}{dx} \left(\frac{\partial g}{\partial y'} \right) \right) dx = 0,$$

and solving each equation for a/b gives

$$-\frac{\int_{x_0}^{x_1} \xi \left(\frac{\partial g}{\partial y} - \frac{d}{dx}\left(\frac{\partial g}{\partial y'}\right)\right) dx}{\int_{x_0}^{x_1} \eta \left(\frac{\partial g}{\partial y} - \frac{d}{dx}\left(\frac{\partial g}{\partial y'}\right)\right) dx} = \frac{a}{b} = -\frac{\int_{x_0}^{x_1} \xi \left(\frac{\partial f}{\partial y} - \frac{d}{dx}\left(\frac{\partial f}{\partial y'}\right)\right) dx}{\int_{x_0}^{x_1} \eta \left(\frac{\partial f}{\partial y} - \frac{d}{dx}\left(\frac{\partial f}{\partial y'}\right)\right) dx}.$$

Upon rearranging we obtain

$$\frac{\int_{x_0}^{x_1} \xi \left(\frac{\partial f}{\partial y} - \frac{d}{dx}\left(\frac{\partial f}{\partial y'}\right)\right) dx}{\int_{x_0}^{x_1} \xi \left(\frac{\partial g}{\partial y} - \frac{d}{dx}\left(\frac{\partial g}{\partial y'}\right)\right) dx} = \frac{\int_{x_0}^{x_1} \eta \left(\frac{\partial f}{\partial y} - \frac{d}{dx}\left(\frac{\partial f}{\partial y'}\right)\right) dx}{\int_{x_0}^{x_1} \eta \left(\frac{\partial g}{\partial y} - \frac{d}{dx}\left(\frac{\partial g}{\partial y'}\right)\right) dx}.$$

The left-hand side is a function of ξ while the right is a function of η, so the only way for these expressions to be identical is for both expressions to be equal to the same constant λ (*the Lagrange multiplier*). Simplifying the equation

$$\frac{\int_{x_0}^{x_1} \eta \left(\frac{\partial f}{\partial y} - \frac{d}{dx}\left(\frac{\partial f}{\partial y'}\right)\right) dx}{\int_{x_0}^{x_1} \eta \left(\frac{\partial g}{\partial y} - \frac{d}{dx}\left(\frac{\partial g}{\partial y'}\right)\right) dx} = \lambda$$

gives

$$\int_{x_0}^{x_1} \eta \left[\frac{\partial(f - \lambda g)}{\partial y} - \frac{d}{dx}\left(\frac{\partial(f - \lambda g)}{\partial y'}\right)\right] dx = 0.$$

Now the usual argument may be continued. Because η is arbitrary, the previous equation can hold only if the term in brackets vanishes. We thus obtain the following *Euler-Lagrange necessary condition* for the constrained problem.

Theorem 4.5.1. *If $y = y(x)$ is a solution to the standard constrained problem, then*

$$\frac{\partial(f - \lambda g)}{\partial y} - \frac{d}{dx}\left(\frac{\partial(f - \lambda g)}{\partial y'}\right) = 0.$$

Example 4.5.2. The *bending energy* of a plane curve is defined to be the integral of squared curvature over the length of the curve, $\int \kappa^2(s)\, ds$. Recall ([**Opr97**], for example) that the curvature of a plane curve also is given by $\dot{\theta} = d\theta/ds$, where $\theta(s)$ is the angle the unit tangent $T(s)$ makes with the x-axis. Physically, a wire will try to assume a shape which minimizes bending energy. (For a biological application of this idea see [**Can70**], as well as [**DH76a**] and [**DH76b**].) Therefore, we can ask the following question. What shape will a wire take if the total turning of its tangent is fixed and the turning is zero at the endpoints? As a constrained variational problem, we write

$$\text{Minimize } J = \int_0^1 \frac{1}{2}\dot\theta^2\,ds \qquad \text{subject to} \qquad I = \int_0^1 \theta\,ds = \frac{1}{6}$$

with $\theta(0) = 0$ and $\theta(1) = 0$. The factor $1/2$ in the first integral is only put in to make calculations cleaner. Minimizing bending energy is clearly equivalent to minimizing one-half of bending energy. Similarly, the $1/6$ value of the second integral is chosen only to avoid fractions later on. The Euler-Lagrange equation for the constrained problem gives

$$\frac{\partial}{\partial\theta}\left(\frac{1}{2}\dot\theta^2 - \lambda\theta\right) - \frac{d}{ds}\left(\frac{\partial}{\partial\dot\theta}\left(\frac{1}{2}\dot\theta^2 - \lambda\theta\right)\right) = 0$$

$$-\lambda - \frac{d}{ds}(\dot\theta) = 0$$

$$\ddot\theta = -\lambda$$

$$\dot\theta = -\lambda s + c$$

$$\theta(s) = -\frac{\lambda s^2}{2} + cs + d.$$

The conditions $\theta(0) = 0$ and $\theta(1) = 0$ give $d = 0$ and then $c = \lambda/2$. Hence

$$\theta(s) = \frac{\lambda}{2}\left(-s^2 + s\right).$$

Now put this $\theta(s)$ into the constraint integral I to obtain

$$\frac{1}{6} = \int_0^1 \theta(s)\,ds$$

$$= \frac{\lambda}{2}\left(-\frac{s^3}{3} + \frac{s^2}{2}\right)\Big|_0^1$$

$$= \frac{\lambda}{2} \cdot \frac{1}{6},$$

giving $\lambda/2 = 1$. Hence, $\theta(s) = -s^2 + s$. A plane curve $\beta(s)$ may be reconstructed from its curvature (see [**Opr97**], for instance, where Maple is used to reconstruct plane and space curves from curvature and torsion). The formula which gives this reconstruction, applied to the bending energy curve, is

$$\beta(s) = \left(\int_0^s \cos(-u^2 + u)\, du, \ \int_0^s \sin(-u^2 + u)\, du \right).$$

This said, a Maple approach to these calculations and a picture of the solution curve $\beta(s)$ may be seen in Example 5.11.6.

Exercise 4.5.3. Solve the constrained problem

$$\text{Extremize} \ \int_0^1 xy\, dx \qquad \text{subject to} \qquad \int_0^1 y'^2\, dx = 1$$

with endpoint conditions $y(0) = 0$ and $y(1) = 0$.

Exercise 4.5.4. Solve the constrained problem

$$\text{Extremize} \ J = \int_0^1 y^2\, dx \qquad \text{subject to} \qquad I = \int_0^1 y'^2\, dx = 1$$

with endpoint conditions $y(0) = 0$ and $y(1) = 0$. Hints: From the Euler-Lagrange equation you will get two cases: $\lambda > 0$ and $\lambda < 0$. Show that the second can't happen. Don't forget to use the endpoint conditions. For the first case, show that $\lambda = 1/n^2\pi^2$ for positive integers n. Use the integral constraint to then show that

$$y(x) = \frac{\sqrt{2}}{n\pi} \sin(n\pi x).$$

Example 4.5.5 (The Catenary). We can now show that a freely hanging chain of fixed length hangs in the shape of a catenary

$$y = c \cosh(x/c + d).$$

The potential energy of the hanging chain is proportional to

$$\int_{x_0}^{x_1} y\sqrt{1 + y'^2}\, dx.$$

This may be seen as follows. Potential energy is given by mgh, and the weight mg of a small piece of chain is given by σs, where σ is a weight density in units of weight per length and $s = \sqrt{1 + y'^2}$ is a small piece of arclength. Of course, the height h is really y, so adding all the small contributions of potential energy, we obtain

$$\text{Potential Energy} = \sigma \int_{x_0}^{x_1} y\sqrt{1 + y'^2}\, dx.$$

The equilibrium position of the chain is assumed when potential energy is minimized. The problem is constrained, however, by the fixed length of the chain $L = \int_{x_0}^{x_1} \sqrt{1 + y'^2}\, dx$. Therefore, in order to see what the equilibrium shape y can be, we extremize

$$J = \int_{x_0}^{x_1} y\sqrt{1 + y'^2} - \lambda\sqrt{1 + y'^2}\, dx.$$

We extremize this integral noting that the integrand is independent of x. The first integral $f - y'(\partial f/\partial y') = c$ gives

$$y\sqrt{1 + y'^2} - \lambda\sqrt{1 + y'^2} - y'\frac{yy'}{\sqrt{1 + y'^2}} + \frac{\lambda y'^2}{\sqrt{1 + y'^2}} = c$$

$$y(1 + y'^2) - \lambda(1 + y'^2) - yy'^2 + \lambda y'^2 = c\sqrt{1 + y'^2}$$

$$y - \lambda = c\sqrt{1 + y'^2}$$

$$(y - \lambda)^2/c^2 = 1 + y'^2$$

$$y' = \frac{\sqrt{(y - \lambda)^2 - c^2}}{c}$$

$$\frac{1}{\sqrt{(y - \lambda)^2 - c^2}}\, dy = \frac{1}{c}\, dx$$

$$\int \frac{1}{\sqrt{(y - \lambda)^2 - c^2}}\, dy = \int \frac{1}{c}\, dx.$$

We can substitute $y - \lambda = c\cosh(w)$ with $dy = c\sinh(w)\, dw$ and integrate (replacing w) to get

$$\text{arccosh}((y - \lambda)/c) = \frac{1}{c}x + d$$

$$y = c\cosh(x/c + d) + \lambda.$$

The d simply shifts the cosh right or left, so by choosing the origin appropriately, we can get $y = c\cosh(x/c) + \lambda$. Of course, λ

adjusts the cosh up or down, so is rather inessential to our general discussion. Therefore, the chain assumes the shape of the catenary $y = c \cosh(x/c)$. Isn't it interesting that the hanging chain is also the profile curve for the minimal surface of revolution, the catenoid?

Exercise 4.5.6 ([Tro96]). Here is a constrained physical problem to try. A circular column of a fluid is rotated about its vertical axis (y-axis) at constant angular speed ω. The upper (free) surface of the fluid assumes a shape which preserves the volume of the fluid and which minimizes the potential energy

$$U(y) = \rho \pi \int_0^R gxy^2 - \omega^2 x^3 y \; dx.$$

Here, R is the radius of the cylinder of fluid, ρ is the mass density of the fluid, g is the gravitational constant and $y(x)$ is the height of the fluid at the radial distance x from the center. The volume of the fluid is

$$V = 2\pi \int_0^R xy \; dx.$$

Show that the cross section $y(x)$ of the upper surface of the fluid is given by

$$y(x) = \frac{\omega^2 x^2}{2g} + \frac{V}{\pi R^2} - \frac{\omega^2 R^2}{4g}.$$

Note that your answer depends on the angular speed, but *not* on the mass density of the fluid. What does this mean? Explain.

Example 4.5.7 (The Isoperimetric Problem). Recall that we suppose we have a closed curve of fixed length $L = \int \sqrt{\dot{x}^2 + \dot{y}^2} \; dt$, and we wish to arrange the curve in a shape which encloses the *maximum* amount of area. Green's theorem allows us to express area as a line integral $\int (1/2)(-y\dot{x} + x\dot{y}) \; dt$, so that the variational problem becomes

$$\text{Extremize} \quad J = \int (1/2)(-y\dot{x} + x\dot{y}) - \lambda\sqrt{\dot{x}^2 + \dot{y}^2} \; dt.$$

The two-variable Euler-Lagrange equations are

$$\frac{\dot{y}}{2} - \frac{d}{dt}\left(-\frac{y}{2} - \frac{\lambda \dot{x}}{\sqrt{\dot{x}^2 + \dot{y}^2}} \right) = 0, \quad -\frac{\dot{x}}{2} - \frac{d}{dt}\left(\frac{x}{2} - \frac{\lambda \dot{y}}{\sqrt{\dot{x}^2 + \dot{y}^2}} \right) = 0,$$

which may be rewritten as

$$\frac{d}{dt}\left(y + \frac{\lambda\dot{x}}{\sqrt{\dot{x}^2 + \dot{y}^2}}\right) = 0, \qquad \frac{d}{dt}\left(-x + \frac{\lambda\dot{y}}{\sqrt{\dot{x}^2 + \dot{y}^2}}\right) = 0.$$

We then have

$$y + \frac{\lambda\dot{x}}{\sqrt{\dot{x}^2 + \dot{y}^2}} = C \quad \text{and} \quad -x + \frac{\lambda\dot{y}}{\sqrt{\dot{x}^2 + \dot{y}^2}} = -D,$$

and a little algebra produces

$$(x - D)^2 + (y - C)^2 = \lambda^2,$$

a circle of radius $r = \lambda$ centered at (D, C). Because, for a circle, $L = 2\pi r$, we have $r = \lambda = L/2\pi$. This calculation is then the explanation behind the loop on a soap film experiment in § 1.3.8. Because the soap film minimizes its surface area and the area inside the ring is fixed, it must be the case that the thread is arranging itself to maximize its enclosed area. By the result derived here, the thread must take the shape of a circle.

Exercise 4.5.8. Take the two-variable Euler-Lagrange equations

$$\frac{\dot{y}}{2} - \frac{d}{dt}\left(-\frac{y}{2} - \frac{\lambda\dot{x}}{\sqrt{\dot{x}^2 + \dot{y}^2}}\right) = 0, \quad -\frac{\dot{x}}{2} - \frac{d}{dt}\left(\frac{x}{2} - \frac{\lambda\dot{y}}{\sqrt{\dot{x}^2 + \dot{y}^2}}\right) = 0,$$

carry out the differentiations with respect to t, and solve for λ, to get

$$-\frac{1}{\lambda} = \frac{\dot{x}\ddot{y} - \ddot{x}\dot{y}}{(\dot{x}^2 + \dot{y}^2)^{3/2}}.$$

The absolute value of the expression on the right is the curvature of the resulting curve, and the equation says that this curvature is constant. A plane curve with constant curvature is a circle (see [**Opr97**, Theorem 1.3.3]).

Exercise 4.5.9. Show by variational methods that the circle also solves the problem of finding the curve of shortest length which encloses a fixed area (see Theorem 1.7.3).

Example 4.5.10 (Bubbles Again). Suppose we ask (as we did in Example 3.11.4) to minimize surface area subject to having a fixed volume. The calculations in Example 3.11.4 give

$$V'(0) = \iint f \, dA \quad \text{and} \quad A'(0) = -2 \iint fH \, dA,$$

where V is volume inside a closed surface and A is the surface area of the surface. According to the Lagrange multiplier method, a solution for our problem is given by solving the Euler-Lagrange equation for

$$J = A - \lambda V,$$

where A and V are given by the usual integrals. Taking a variation with respect to a parameter t and assuming the minimum occurs at $t = 0$ gives $J'(0) = 0$ and

$$0 = A'(0) - \lambda V'(0)$$
$$= \iint -2 fH - \lambda f \, dA$$
$$= \iint (\lambda + 2H) f \, dA.$$

Since this is true for all f, it must be the case that $\lambda + 2H = 0$. Therefore, $H = -\lambda/2$, and mean curvature is constant. This provides another proof of Theorem 3.11.5 and says that soap bubbles have constant mean curvature.

Exercise 4.5.11. Analyze the problem of maximizing volume subject to a fixed surface area. What does this tell you in general about constrained problems such as we have been considering? Are the roles of the quantity to be minimized or maximized and the constrained quantity always interchangeable?

Exercise 4.5.12 (Surfaces of Delaunay). In this exercise, we specialize Example 4.5.10 to the case of a surface of revolution to see if we can get more information then. Of course, we know by Example 4.5.10 that these surfaces have constant mean curvature, so they are surfaces of Delaunay (see Exercise 3.11.2). Here we derive that fact directly from volume-surface area considerations. Therefore, suppose we have a surface of revolution which encloses a fixed volume

$\mathcal{V} = \pi \int h(u)^2 \, du$ such that surface area $\mathcal{S} = 2\pi \int h(u)\sqrt{1 + h'(u)^2} \, du$ is minimized. What are the resulting surfaces? Neglecting π in the formula, the set-up for this constrained problem takes the form

$$\text{Extremize } J = \int 2h(u)\sqrt{1 + h'(u)^2} - \lambda h(u)^2 \, du.$$

Show that the extremals for this constrained problem are the surfaces of Delaunay (see Exercise 3.11.2). Therefore, we should expect to see surfaces of Delaunay in Nature where surface area is minimized subject to a fixed volume. In fact, if you look in [**Tho92**], you will see one-celled creatures with just the right shapes (those in § 5.13).

Hints: Since the integrand does not depend on the independent variable u, you may use the first integral $f - y'(\partial f / \partial y') = c$ in place of the Euler-Lagrange equation to get

$$2h\sqrt{1 + h'^2} - \lambda h^2 - h'\left(\frac{2hh'}{\sqrt{1 + h'^2}}\right) = c.$$

Show that you get

$$h^2 \pm \frac{2ah}{\sqrt{1 + h'^2}} = \pm b^2,$$

where $a = -1/\lambda$ and $b = -c/\lambda$. Compare with Exercise 3.11.2 and see § 5.13 for a Maple approach.

Example 4.5.13 (The Mylar Balloon [**Pau94**]). The calculus of variations may be used to find the shape of mylar balloons often found at birthday parties. Here we will describe the variational set-up and find solutions in terms of integrals. In § 5.14, Maple will be used to calculate the integrals in question and to draw the predicted shape of the balloon. A mylar balloon is constructed by taking two disks of mylar, sewing them along their boundaries and inflating. We can think of the top half of the inflated balloon mathematically as a curve $y = y(x)$ revolved about the y-axis. The bottom half is just a reflection of the top through the xz-plane. So, we can write a parametrization for the top as $\mathbf{x}(u, v) = (u\cos(v), y(u), u\sin(u))$, since the profile curve is $\alpha(u) = (u, y(u), 0)$ and we revolve around the y-axis. Note that, in § 5.14, we revolve around the z-axis to accommodate Maple.

Suppose the disks have radius r. Mylar doesn't stretch appreciably, so the profile curve α must also have length r from the top of the balloon to where it hits the x-axis, at a say. This a is called the *inflated radius* of the balloon. Then we must have

$$\int_0^a \sqrt{1 + y'^2}\, dx = r,$$

since the integral on the left gives the arclength of the profile curve from 0 to a. This looks suspiciously like a constraint in a variational problem, but what is the integral to be extremized? Remember, the balloon is inflated to its final shape, so it seems likely that the pressure of the enclosed gas (e.g. air, helium) is forcing the shape to maximize volume. Why is this problem different from the problem of maximizing volume subject to fixed surface area? If it were not different, then the balloon would be a sphere. Since mylar balloons are definitely not spheres, something must be wrong about the surface area interpretation. What goes wrong can be observed in the wrinkling or crimping along the sewn equator of the inflated balloon. Because the mylar does not want to stretch, crimping occurs along the sewn equator and we see that it is not surface area which constrains the problem, but rather the length of the profile curve. The volume of the surface of revolution is given by

$$V = 2 \cdot 2\pi \int_0^a xy\, dx,$$

where the extra 2 comes from including the bottom half of the balloon. The problem is then to extremize

$$4\pi \int_0^a xy\, dx \quad \text{subject to} \quad \int_0^a \sqrt{1 + y'^2}\, dx = r.$$

The constrained Euler-Lagrange equation then gives

$$4\pi x - \frac{d}{dx}\left(-\frac{\lambda y'}{\sqrt{1 + y'^2}} \right) = 0$$

$$\int \frac{d}{dx}\left(-\frac{\lambda y'}{\sqrt{1 + y'^2}} \right) dx = \int 4\pi x\, dx$$

$$-\frac{\lambda y'}{\sqrt{1 + y'^2}} = 2\pi x^2 + c.$$

Now, there are two assumptions about the shape of the balloon which we will make. A general principle in variational problems is that to maximize volume, the profile curve must meet the axes at $x = 0$ and $x = a$ perpendicularly. Intuitively, we can see this by symmetry considerations. Analytically, this means that $y'(0) = 0$ and, as x approaches a, $y'(x)$ approaches $-\infty$. The minus sign arises because the profile curve decreases for $0 < x < a$, so $y'(x) < 0$ there. At $x = 0$, therefore, we have $y' = 0$, so we have $c = 0$ above. Also, because $y'(x) < 0$, the equation above implies that $\lambda > 0$. Thus, let $\lambda = 2\pi\,k^2$ for some k. Then

$$-\frac{\lambda y'}{\sqrt{1 + y'^{\,2}}} = 2\pi\,x^2$$

$$-\frac{y'}{\sqrt{1 + y'^{\,2}}} = -\frac{x^2}{k^2}$$

$$y' = -\frac{x^2}{\sqrt{k^4 - x^4}}$$

$$y(x) = \int_x^a \frac{t^2}{\sqrt{k^4 - t^4}}\,dt.$$

The upper limit of integration comes from the requirement that $y(a) = 0$. We also need to know what k can be. To see this, use the fact that $y' \mapsto -\infty$ as $x \mapsto a$. This will happen when $k = a$. Therefore, we have determined the profile curve of the mylar balloon to be

$$y(x) = \int_x^a \frac{t^2}{\sqrt{a^4 - t^4}}\,dt.$$

Knowing this, it is an easy matter to calculate y' and

$$r = \int_0^a \sqrt{1 + y'^{\,2}}\,dx$$

$$= \int_0^a \frac{a^2}{\sqrt{a^4 - t^4}}\,dt.$$

This integral can be simplified to allow for numerical evaluation by making the substitutions

$$x = at^{1/4}, \qquad dx = \frac{1}{4}\,at^{-3/4}\,dt.$$

Then we obtain

$$r = \frac{a}{4} \int_0^1 \frac{t^{-3/4}}{\sqrt{1-t}} \, dt.$$

This then relates the radii of the disks to the inflated radius of the balloon. Similarly, the thickness of the balloon is given by $2 \cdot y(0)$, so

$$\text{thickness} = 2 \int_0^a \frac{x^2}{\sqrt{a^4 - x^4}} \, dx$$

$$= \frac{a}{2} \int_0^1 \frac{t^{-1/4}}{\sqrt{1-t}} \, dt,$$

using the same substitution $x = at^{1/4}$. Finally, the volume of the mylar balloon may be calculated:

$$V = 4\pi \int_0^a xy \, dx$$

$$= 4\pi \int_0^a \int_x^a \frac{xt^2}{\sqrt{a^4 - t^4}} \, dt \, dx$$

$$= 4\pi \int_0^a \int_0^t \frac{xt^2}{\sqrt{a^4 - t^4}} \, dx \, dt$$

$$= 2\pi \int_0^a \frac{t^4}{\sqrt{a^4 - t^4}} \, dt$$

by changing the order of integration. (The reader should verify that both sets of integration limits give the region under the diagonal $x = t$ in the square $[0, a] \times [0, a]$.) Again, using a substitution $t = au^{1/4}$, we get

$$V = \frac{\pi a^3}{2} \int_0^1 \frac{u^{1/4}}{\sqrt{1-u}} \, du.$$

Both the thickness and the volume may be related to the original disk radius r by plugging in the equation relating a and r. This is done in § 5.14, where a picture of the balloon awaits.

Chapter 5

Maple, Soap Films, and Minimal Surfaces

5.1. Introduction

In this chapter, we will use Maple to solve problems and illustrate concepts from the previous chapters. This will include some of the topics from Chapter 1, such as capillarity, as well as various topics from minimal surface theory. The point must be made that Maple doesn't simply plot parametrizations to give nice pictures, but provides the power to do things such as solving differential equations numerically to allow for visualization. Also, we can analyze problems such as when catenoids or pieces of Enneper's surface minimize area with Maple. While rigorous mathematical analysis of some subjects such as this may be too advanced for undergraduates, here we see that Maple analysis is most convincing and, what's more, comprehensible. The procedures below have been tested in MapleV version 4, MapleV version 5 and Maple 6. There are some differences in how they work in the three incarnations of Maple, and these differences are noted.

5.2. Fused Bubbles

Recall from Chapter 1 that there is a relationship among the radii of fused bubbles (see Example 1.5.5). The following procedure displays

a cross-section of two fused bubbles. The procedure creates the two
fused bubbles and their interface just from knowing the radius r_A of
the large bubble (assumed centered at the origin) and the angle ϕ
which the vector r_A from the origin to the intersection point makes
with the x-axis. Let β be the angle (inside the triangle with ϕ) which
the vector from B to the intersection point makes with the x-axis and
γ the angle which the vector from C to the intersection point makes
with the x-axis.

Exercise 5.2.1. Choose various inputs for the procedure (as shown
in the examples below) and then answer the following questions.

(1) Show that $\beta = 120 - \phi$ and $\gamma = 60 - \phi$. What conditions
does this place on ϕ?

(2) After working through the procedures below, make a *conjecture* about where the point B can be located. Verify this
conjecture. (Hint: the Law of Cosines and an estimate may
come in handy.)

In the procedure below, α is the parametrized line through B and the
intersection point of the bubbles. Note that we are using Plateau's
120° rule here. Since B lies on the x-axis, T finds this point and substitution produces B. The same approach (with a different angle, of
course) is used to find C (where β in the procedure is the parametrized
line through the intersection point and C), while r_C comes from solving the reciprocal relation in Example 1.5.5. In this description, we
shall use r_A, etc., for both the vector and distance when no confusion
will arise. The plots of the bubbles around A and B are just complete circles, while the interface between the two intersection points
must be plotted only inside the circles. This requires determining the
angle τ (called 'angle' in the procedure) between r_C and $-C$ and the
relation $r_C \cdot (-C) = |r_C| \cdot |-C| \cos(\tau)$ with $r_C = r_A - C$. We solve for
the angle by using arccos. Note that we use our own version of the
dot product (called 'dp2' below). This is because, while the simple
command 'dotprod' works well in MapleV version 4, it does not in
MapleV version 5. The 'dotprod' command in version 5 and in Maple
6 appears to give a general hermitian inner product which does not

allow for easy real simplification. Although the extra argument 'dot-prod(,orthogonal)' may provide the real dot product, but it is much easier to write a 'dp' procedure rather than wade through Maple's list of extras.

```
>  with(plots):with(linalg):
>  dp2:= proc(X,Y)
simplify(X[1]*Y[1] + X[2]*Y[2]);
end:
>  fusedbub:=proc(r_A,phi)
local alpha,T,B,r_B,r_C,beta,T2,C,plot1,plot2,
angle,plot3;
alpha:=evalm([r_A*cos(phi),r_A*sin(phi)]+
[t*cos(phi-2*Pi/3), t*sin(phi-2*Pi/3)]);
T:=fsolve(alpha[2]=0,t);
B:=evalf([subs(t=T,alpha[1]),0]);
r_B:=evalf(norm(evalm(B-[r_A*cos(phi),
r_A*sin(phi)]),2));
r_C:=fsolve(1/r_B=1/r_A + 1/x,x);print('r_A=',r_A,
'r_B=',r_B,'r_C=',r_C);
beta:=evalm([r_A*cos(phi),r_A*sin(phi)]+
[t*cos(phi-Pi/3), t*sin(phi-Pi/3)]);
T2:=fsolve(beta[2]=0,t);
C:=evalf([subs(t=T2,beta[1]),0]);
plot1:=plot([r_A*cos(theta),r_A*sin(theta),
theta=0..2*Pi],color=red):
plot2:=plot([B[1]+r_B*cos(theta),r_B*sin(theta),
theta=0..2*Pi],color=blue):
angle:=arccos(dp2([r_A*cos(phi)-C[1],r_A*sin(phi)],
-C)/(norm([r_A*cos(phi)-C[1],
r_A*sin(phi)],2)*C[1]));
plot3:=plot([C[1]+r_C*cos(theta),r_C*sin(theta),
theta=Pi-angle..Pi+angle],color=black):
print('A=',[0,0],'B=',B,'C=',C);
display({plot1,plot2,plot3},scaling=constrained,
view=[-r_A..B[1]+r_B,-r_A..r_A]):
end:
>  fusedbub(1,Pi/4);
```

$$r_A = 1, \quad r_B = .7320508075, \quad r_C = 2.732050806,$$

$$A = [0, 0], \quad B = [.8965754718, 0], \quad C = [3.346065215, 0]$$

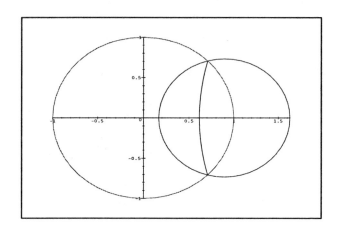

```
>   fusedbub(1,Pi/3-.0001);
```

$$r_A = 1, \quad r_B = .9998845366, \quad r_C = 8659.733107,$$

$$A = [0, 0], \quad B = [.9999422736, 0], \quad C = [8660.254053, 0]$$

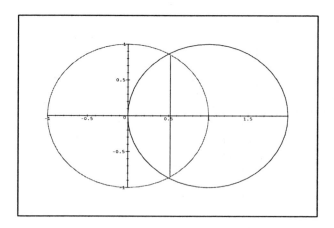

```
>   fusedbub(1,.07);
```

$$r_A = 1, \quad r_B = .07781145860, \quad r_C = .08437695234,$$

$$A = [0, 0], \quad B = [.9634537688, 0], \quad C = [1.044747062, 0]$$

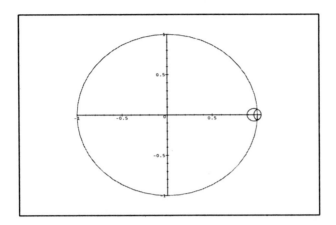

The following is another Maple procedure (due to J. Reinmann) which plots fused bubbles given the radius of the largest (i.e. r_A) and the angle ϕ which r_A makes with the x-axis.

Exercise 5.2.2. Match up the ingredients of the procedure below with the discussion in Example 1.5.5. If you don't get things at first, leave this and come back to it after you have gone through more Maple things below.

```
> with(plots):
> fusedbub2:=proc(rA,phi) #phi in radians
local AB,BC,rB,rC,circA,circB,circC;
rB:=rA*sin(phi)/sin(2*Pi/3-phi);
rC:=rA*rB/(rA-rB);
AB:=rA*sin(Pi/3)/sin(2*Pi/3-phi);
BC:=rC*sin(Pi/3)/sin(Pi/3+phi);
circA:=implicitplot(x^2+y^2=rA^2, x=-rA..rA*cos(phi),
y=-rA..rA, color=black):
circB:=implicitplot((x-AB)^2+y^2=rB^2, x=rA*cos(phi)..
AB+rB, y=-rB..rB,color=black):
circC:=implicitplot((x-AB-BC)^2+y^2=rC^2, x=AB+BC-rC..
rA*cos(phi),y=-rA*sin(phi)..rA*sin(phi),color=black):
display({circA,circB,circC},scaling=constrained);
end:
```

> `fusedbub2(1,Pi/4);`

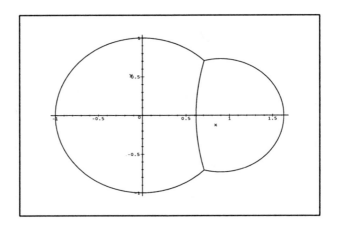

> `fusedbub2(1,Pi/3-0.0001);`

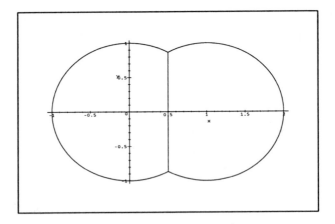

> fusedbub2(1,.07);

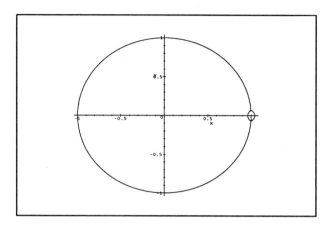

5.3. Capillarity: Inclined Planes

5.3.1. Maple and the Differential Equation. This section provides a procedure (which is a modified version of one created by John Reinmann) for plotting the surface shape of a liquid in contact with an inclined plane (see Example 1.6.1) . First, let's use Maple to solve the differential equations obtained in Chapter 1, Example 1.6.1 and Exercise 1.6.2. The first equation is for fluid rise on the left side of the inclined plane and the second equation is for fluid rise on the right:

```
>  dsolve(diff(y(x),x)=2*y(x)*sqrt(1-y(x)^2)/ (1-
   2*y(x)^2), y(x));
```

$$\frac{1}{2}\operatorname{arctanh}\left(\frac{1}{\sqrt{1-\mathrm{y}(x)^2}}\right) - \sqrt{1-\mathrm{y}(x)^2} + x = {\_C1}$$

```
>  dsolve(diff(y(x),x)=-2*y(x)*sqrt(1-y(x)^2)/( 1-
   2*y(x)^2), y(x));
```

$$-\frac{1}{2}\operatorname{arctanh}\left(\frac{1}{\sqrt{1-\mathrm{y}(x)^2}}\right) + \sqrt{1-\mathrm{y}(x)^2} + x = {\_C1}$$

5.3.2. The Procedure. Now we can give the procedure which plots the fluid rise on an inclined plane. Notice that the two inputs of the procedure are the contact angle of the fluid with the plane and the inclination of the plane. This looks like a rather complicated procedure at first, but the basic idea is to plot three things separately: the left side of the fluid, the plane (cross-section), and the right side of the fluid. To plot the fluid, the solutions of the differential equations above are used with different initial conditions determining the constant. (Compare the exposition in Example 1.6.1.) Namely, the height of the left fluid rise, denoted $h1$ in the procedure below, is related to the x-coordinate below it by $x = h1\cot(b)$ (where $b = \beta$, the angle of inclination of the plane). Plugging these in the first equation above gives $C1 = 1/2\operatorname{arctanh}(1/\sqrt{1-h1^2}) - \sqrt{1-h1^2} + h1\cot(b)$, and, in turn, this result may be plugged in the general equation to give eq1 below. Also note that, by symmetry, we can restrict ourselves to the case where the angle of inclination of the plane has $0° \le b \le 90°$ and the contact angle has $0° \le a \le 180°$. To make things easy, the input angles are in degrees and are converted to radians for Maple's benefit. Finally, note that $a + b = 180°$ and $a = b$ are not allowed for input. Try one of these forbidden conditions to see what happens. What do these conditions mean physically?

```
>  with(plots):

>  liquid_rise:=proc(ContactAngle,InclineAngle)
local a,b,d,h1,h2,hmax,eq1,eq2,k1,k2,left,right,
plane,xhi1,xlo2,xmax,ylo1,ylo2,yhi1,yhi2;
a:=evalf(ContactAngle*Pi/180);
b:=evalf(InclineAngle*Pi/180);
h1:=evalf(sin((b-a)/2));
h2:=evalf(sin((Pi-(a+b))/2));
hmax:=max(abs(h1),abs(h2));
eq1:=x=sqrt(1-y^2)-sqrt(1-h1^2)-arctanh(1/sqrt(1-
y^2))/2 + arctanh(1/sqrt(1-h1^2))/2+h1*cot(b);
k1:=evalf(rhs(subs(y=h1,eq1)));
if h1>=0 then ylo1:=0;yhi1:=h1;xhi1:=k1;
else ylo1:=h1;yhi1:=0;xhi1:=0;fi;
eq2:=x=-sqrt(1-y^2)+sqrt(1-h2^2)+arctanh(1/sqrt(1-
y^2))/2 - arctanh(1/sqrt(1-h2^2))/2+h2*cot(b);
k2:=evalf(rhs(subs(y=h2,eq2)));
if h2>=0 then ylo2:=0;yhi2:=h2;xlo2:=0;
else ylo2:=h2;yhi2:=0;xlo2:=k2;fi;
xmax:=2*max(abs(k1),abs(k2),abs(h1),abs(h2));
left:=implicitplot(eq1,x=-xmax..xhi1,y=ylo1..yhi1,
color=blue);
right:=implicitplot(eq2,x=xlo2..xmax,y=ylo2..yhi2,
color=red);
plane:=plot([cot(b)*t,t,t=-hmax..hmax],thickness=2,
color=black);
display({left,right,plane},scaling=constrained,
xtickmarks=0,ytickmarks=0);
end:
```

> `liquid_rise(10,60);`

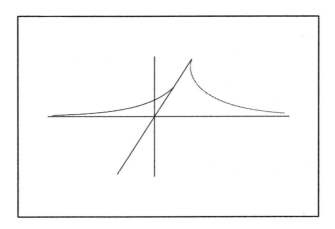

> `liquid_rise(0,90);`

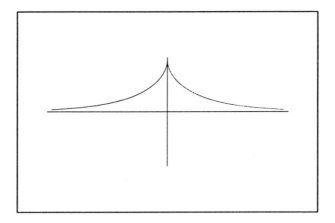

For the next four cases, note what happens when we keep the contact angle constant, but raise the angle of inclination of the plane. When the angle of inclination equals the contact angle, what do you predict? Why?

```
>   liquid_rise(70,50);
```

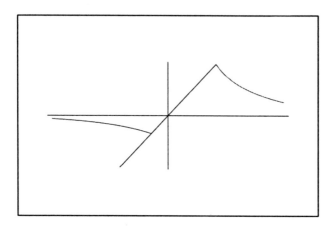

```
>   liquid_rise(70,60);
```

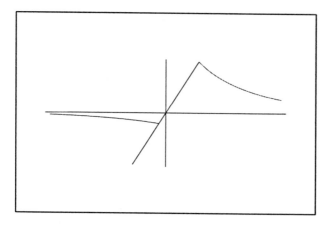

```
> liquid_rise(70,80);
```

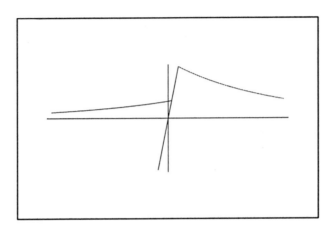

```
> liquid_rise(70,85);
```

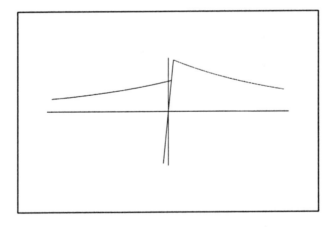

```
>  liquid_rise(90,45);
```

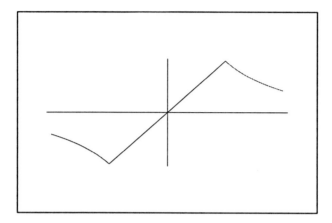

Remember that sometimes liquids make contact 'the wrong way'. For example, mercury makes contact with glass at an angle of 140°. Look what this means for the inclined plane.

```
>  liquid_rise(150,80);
```

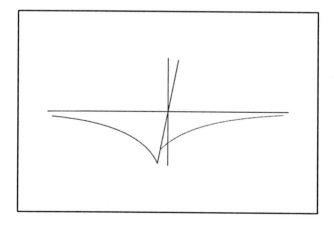

5.4. Capillarity: Thin Tubes

Now let's consider the situation of a liquid in a thin tube (see Example 1.6.5). The following procedure uses the results of Example 1.6.5 to plot the surface of the liquid. Note that, if the contact angle α has $0° \leq \alpha \leq 90°$, then liquid rises in the tube, while, if $90° < \alpha \leq 180°$, then the liquid falls in the tube. Note that the inputs of the procedure are the surface tension σ, the tube radius r and the contact angle α. We take the density $\rho = 1$.

```
>   with(plots):

>   captube:=proc(sigma,r,alpha)
local radalpha,k,ybot,R,meniscus,plot1,plot2,
plot3,plot4;
radalpha:=Pi/180*alpha;
k:=2*sigma/980;
ybot:=k*cos(radalpha)/r-r*(sec(radalpha)-
2/3*sec(radalpha)^3+2/3*tan(radalpha)^3);
R:=r/cos(radalpha);
meniscus:=[R*cos(t),R+ybot+R*sin(t)];
plot1:=plot([-r,t,t=0..R+ybot],color=blue,
thickness=2):
plot2:=plot([r,t,t=0..R+ybot],color=blue,
thickness=2):
plot3:=plot({[t,0,t=-3*r..-r],[t,0,t=r..3*r]},
color=blue,thickness=2):
plot4:=plot([meniscus[1],meniscus[2],t=Pi+radalpha..
2*Pi-radalpha],color=red,
thickness=2):
display({plot1,plot2,plot3,plot4},scaling=constrained,
xtickmarks=2,ytickmarks=4,axes=boxed);
end:
```

From the surface tension table of Chapter 1, we see that the following example shows the capillary surface determined by water in a glass tube (where the contact angle is zero).

```
>  captube(73,0.1,0);
```

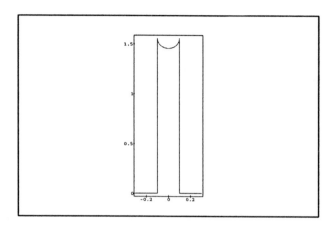

Next we see soap solution in the tube. Notice that the solution doesn't rise as high as plain water, because the surface tension is smaller.

```
>  captube(25,0.1,0);
```

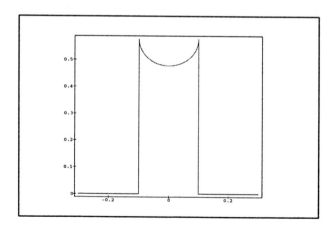

Here is water in a paraffin tube with contact angle 107°. Notice that the level falls.

```
>  captube(73,0.1,107);
```

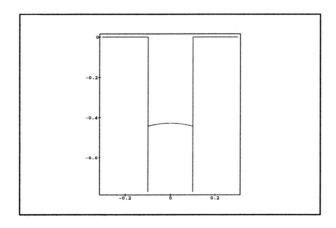

Finally, mercury in a glass tube falls to a much greater degree. In fact, to get a reasonable picture, we increased the width of the tube.

```
>  captube(465,0.5,140);
```

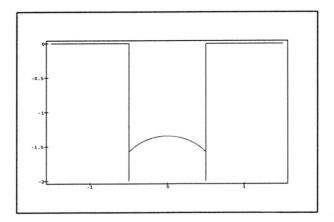

5.5. Minimal Surfaces of Revolution

5.5.1. The Basic Procedures. This section shows how Maple may be used to find minimal surfaces of revolution. Recall that a surface is minimal if its mean curvature vanishes: $H = 0$. We start with the usual procedures needed for finding Gauss and mean curvatures. (Note that we create our own dot product procedure 'dp'. Although Maple's 'dotprod' procedure works fine in MapleV version 4, it was redefined in version 5 to be more general (and less usable). It is possible that version 5's 'dotprod(,orthogonal)' may work, but it is a simpler matter to just create a dot product procedure you know works.)

```
> with(plots):with(linalg):
```

For a parametrization $\mathbf{x}(u, v)$, the next procedures compute $\mathbf{x}_u$, $\mathbf{x}_v$ and the metric $E = \mathbf{x}_u \cdot \mathbf{x}_u$, $F = \mathbf{x}_u \cdot \mathbf{x}_v$ and $G = \mathbf{x}_v \cdot \mathbf{x}_v$:

```
> dp := proc(X,Y)
simplify(X[1]*Y[1] + X[2]*Y[2] + X[3]*Y[3]);
end:
```

```
> EFG := proc(X)
local E,F,G,Xu,Xv;
Xu := [diff(X[1],u),diff(X[2],u),diff(X[3],u)];
Xv := [diff(X[1],v),diff(X[2],v),diff(X[3],v)];
E := dp(Xu,Xu);
F := dp(Xu,Xv);
G := dp(Xv,Xv);
simplify([E,F,G]);
end:
```

The following procedure gives the unit normal to the input surface X. If you wish to take $UN(X)$, then X must be given in parametrized form in square brackets as done down below:

```
> UN := proc(X)
local Z,s,Xu,Xv;
Xu := [diff(X[1],u),diff(X[2],u),diff(X[3],u)];
Xv := [diff(X[1],v),diff(X[2],v),diff(X[3],v)];
Z := crossprod(Xu,Xv);
s := sqrt(Z[1]^2+Z[2]^2+Z[3]^2);
simplify([Z[1]/s,Z[2]/s,Z[3]/s]);
end:
```

The formula for Gauss curvature requires dotting certain second partial derivatives with the unit normal. This is done as follows:

```
> lmn := proc(X)
local Xu,Xv,Xuu,Xuv,Xvv,U,l,m,n;
Xu :=[diff(X[1],u),diff(X[2],u),diff(X[3],u)];
Xv :=[diff(X[1],v),diff(X[2],v),diff(X[3],v)];
Xuu :=[diff(Xu[1],u),diff(Xu[2],u),diff(Xu[3],u)];
Xuv :=[diff(Xu[1],v),diff(Xu[2],v),diff(Xu[3],v)];
Xvv :=[diff(Xv[1],v),diff(Xv[2],v),diff(Xv[3],v)];
U := UN(X);
l := dp(U,Xuu);
m := dp(U,Xuv);
n := dp(U,Xvv);
simplify([l,m,n]);
end:
```

Finally we can calculate Gauss curvature K as follows:

```
> GK := proc(X)
local E,F,G,l,m,n,S,T;
S := EFG(X);
T := lmn(X);
E := S[1];
F := S[2];
G := S[3];
l := T[1];
m := T[2];
n := T[3];
simplify((l*n-m^2)/(E*G-F^2));
end:
```

Mean curvature H is given by

```
> MK := proc(X)
local E,F,G,l,m,n,S,T;
S := EFG(X);
T := lmn(X);
E := S[1];
F := S[2];
G := S[3];
l := T[1];
m := T[2];
n := T[3];
simplify((G*l+E*n-2*F*m)/(2*E*G-2*F^2));
end:
```

Exercise 5.5.1. Compute the mean and Gauss curvatures for the catenoid and helicoid.

5.5.2. Minimal Surface of Revolution. Now we can enter a function h which will be the profile curve of a surface of revolution with parametrization $\mathbf{x}(u, v) = (u, h(u) \cos(v), h(u) \sin(v))$:

```
>  h:=t->h(t);
```

$$h := h$$

```
>  surfrev:=[u,h(u)*cos(v),h(u)*sin(v)];
```

$$\textit{surfrev} := [u,\, \mathrm{h}(u) \cos(v),\, \mathrm{h}(u) \sin(v)]$$

We can take the mean curvature using the MK procedure above:

```
>  mean:=MK(surfrev);
```

$$mean := \frac{1}{2} \frac{-\mathrm{h}(u)\,(\frac{\partial^2}{\partial u^2}\,\mathrm{h}(u)) + (\frac{\partial}{\partial u}\,\mathrm{h}(u))^2 + 1}{\sqrt{\mathrm{h}(u)^2\,(1 + (\frac{\partial}{\partial u}\,\mathrm{h}(u))^2)}\,(1 + (\frac{\partial}{\partial u}\,\mathrm{h}(u))^2)}$$

We want $H = 0$, so we take the numerator of mean and solve the differential equation obtained when we set the numerator equal to zero:

```
>  numer(mean);
```

$$-\mathrm{h}(u)\,(\frac{\partial^2}{\partial u^2}\,\mathrm{h}(u)) + (\frac{\partial}{\partial u}\,\mathrm{h}(u))^2 + 1$$

```
>  dsolve(numer(mean)=0,h(u));
```

$$u = -\frac{\ln(\sqrt{\_C1}\,\mathrm{h}(u) + \sqrt{-1 + \mathrm{h}(u)^2\,\_C1})}{\sqrt{\_C1}} - \_C2,$$

$$u = \frac{\ln(\sqrt{\_C1}\,\mathrm{h}(u) + \sqrt{-1 + \mathrm{h}(u)^2\,\_C1})}{\sqrt{\_C1}} - \_C2$$

We get two answers because we need square roots to get u. We can take one answer as follows and solve for the function $h(u)$:

```
>  sol:=dsolve(numer(mean)=0,h(u))[1];
```

$$sol := u = -\frac{\ln(\sqrt{\_C1}\,\mathrm{h}(u) + \sqrt{-1 + \mathrm{h}(u)^2\,\_C1})}{\sqrt{\_C1}} - \_C2$$

In fact, in Maple 6, the default is for 'dsolve' to solve for the dependent variable, so the output to the 'dsolve' command above actually will have $h(u)$ isolated and will be in exponential notation. This means that the next step is unnecessary (though harmless) in Maple 6:

```
>  ans:=simplify(solve(sol,h(u)));
```

$$ans := \frac{1}{2} \frac{\left(e^{(2\sqrt{-C1}\,(u+\_C2))} + 1\right) e^{(-\sqrt{-C1}\,(u+\_C2))}}{\sqrt{\_C1}}$$

```
>  realans:=combine(convert(ans,trig),trig);
```

$$realans := \frac{\cosh(u\sqrt{\_C1} + \_C2\sqrt{\_C1})}{\sqrt{\_C1}}$$

We can pick _C1 and _C2 to see what this function is in one case. Note that _C1 must be nonzero since it is in the denominator.

```
>  func:=subs({_C1=1,_C2=0},realans);
```

$$func := \cosh(u)$$

If you don't recognize this function yet, you can plot it — and the surface of revolution!

```
>  plot(func,u=-2..2);
```

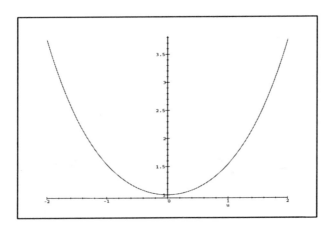

```
>  plot3d([u,func*cos(v),func*sin(v)],u=-2..2, v=0..
2*Pi,scaling=constrained,style=patch,shading=zhue,
lightmodel=light3 );
```

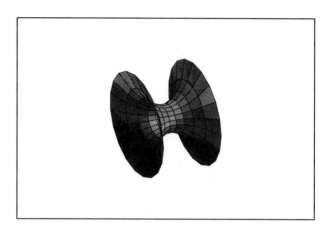

So now we know what a minimal surface of revolution must be. What happens if we weaken the hypothesis a bit? Namely, what if we require the minimal surface to be constructed by circles parallel to the xy-plane, say, but not necessarily concentric about an axis. This problem is discussed in [**Nit89**, §§95-96] with special case solution given by

$$x = \int_b^u \frac{t^2}{\Delta(t)}\, dt + u\cos(v),$$
$$y = u\sin(v),$$
$$z = ab \int_b^u \frac{1}{\Delta(t)}\, dt,$$

where $\Delta(t) = \sqrt{(t^2 + a^2)(t^2 - b^2)}$ for $0 < b \le a$. Here b is the radius of the smallest circle generating the surface (lying in the plane $z = 0$). The centers of the generating circles lie on the curve

$$\left(\int_b^u \frac{t^2}{\Delta(t)}\, dt, 0, ab \int_b^u \frac{1}{\Delta(t)}\, dt \right).$$

Further, the inequality

$$\int_b^u \frac{t^2}{\Delta(t)}\, dt < \int_b^u \frac{t}{\sqrt{t^2 - b^2}}\, dt = \sqrt{u^2 - b^2} < u$$

says that the horizontal distance between two generating circles is less than the sum of their radii. That is, the two circles overlap as seen from above. Here is a Maple approach to graphing these *skew catenoids*:

```
> with(plots):

> skewcat:=proc(a,b,up)
local X,desys,mu,nu,Delta1,p1,p2;
Delta1:=sqrt((u^2+a^2)*(u^2-b^2));
X:=[f+u*cos(v),u*sin(v),a*b*g];
desys:=dsolve({diff(m(u),u)=u^2/Delta1,diff(n(u),u)=
1/Delta1,n(b)=0,m(b)=0},{n(u),m(u)},type=numeric,
output=listprocedure):
mu:=subs(desys,m(u));nu:=subs(desys,n(u));
p1:=plot3d(subs({f='mu'(u),g='nu'(u)},X),u=b..up,
v=0..2*Pi,grid=[12,20]):
p2:=plot3d(subs({f='-mu'(u),g='-nu'(u)},X),u=b..up,
v=0..2*Pi,grid=[12,20]):
display3d({p1,p2},style=patch,shading=xy,scaling=
constrained,orientation=[81,83]);
end:

> skewcat(2,0.4,2);
```

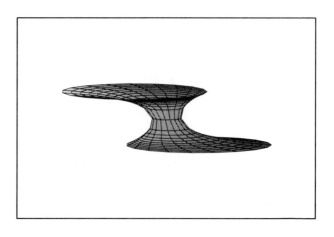

5.6. The Catenoid versus Two Disks

This section shows how Maple may be used to analyze the problem of when the surface area of the catenoid is greater than that of two disks. (Special thanks go to John Reinmann for suggesting ways to do things here, such as non-dimensionalizing the problem.) Recall Exercise 3.3.7

Exercise 5.6.1. Let x_0, $-x_0$ be the points on the x-axis which are centers of given disks of radius y_0. Also consider a catenoid which passes through $(-x_0, y_0)$ and (x_0, y_0).

(1) If $x_0/y_0 > .663$ (approximately), then there is no catenoid joining the points.

(2) If $x_0/y_0 > .528$ (approximately), then the two disks give an absolute minimum for surface area. This is the so-called Goldschmidt discontinuous solution.

(3) If $x_0/y_0 < .528$ (approximately), then a catenoid is the absolute minimum and the Goldschmidt solution is a local minimum.

(4) If $.528 < x_0/y_0 < .663$ (approximately), then the catenoid is only a local minimum.

Of course, we must start with the following basic commands.

```
>    with(plots):with(linalg):
```

5.6.1. Catenaries. We now non-dimensionalize the problem by writing $y = c \cosh(x/c)$ as

$$\frac{x}{y} = \frac{c}{y} \operatorname{arccosh}\left(\frac{y}{c}\right)$$
$$= u \operatorname{arccosh}\left(\frac{1}{u}\right),$$

for $u = c/y$. So let's plot x/y (on the vertical axis) as a function of $u = c/y$.

```
> plot(u*arccosh(1/u),u=0..1);
```

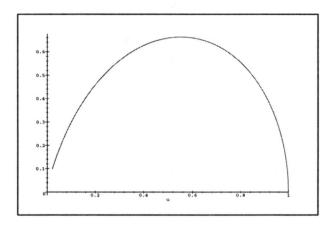

Now we can see what is happening. For each x/y up to some maximum, there are *two* values of c/y which produce a catenary through $(-x, y)$ and (x, y). When we do the two rings soap film experiment (see § 1.3.10), we always see the same catenoid. Why don't we see two? We'll see why below. First, let's find the maximum point on the graph by finding critical points. Maple does this using the 'diff' command to differentiate and the 'fsolve' command to solve when we set the derivative to zero. The 'f' in 'fsolve' stands for 'floating point', so we get a decimal answer. To get the correct answer for the graph above in Maple 6, it is necessary to use the extra option of telling Maple in what interval to look for the solution. That is, in Maple 6, write

```
> fsolve(diff(u*arccosh(1/u),u)=0,u,u=0..1);
```

This also does no harm in versions 4 and 5, but we shall stick to the version 4 command:

```
> fsolve(diff(u*arccosh(1/u),u)=0,u);
```

$$.5524341245$$

Now use the substitution command 'subs' to plug .5524341245 into $u \operatorname{arccosh}(1/u)$ to get the maximum x/y value:

```
>  evalf(subs(u=.5524341245,u*arccosh(1/u)));
```

$$.6627434192$$

This is part (1) of the exercise. For $x/y > .6627434192$, there are *no* catenaries passing through the given points. Again, notice that Maple has rounded here, so the numbers given are approximations (albeit good ones). As examples, let's take $x/y = 0.3$ and find the c's which give the two corresponding catenaries:

```
>  fsolve(.3=u*arccosh(1/u),u,0..0.5524341245);
```

$$.1003411682$$

```
>  fsolve(.3=u*arccosh(1/u),u,.5524341245..1);
```

$$.9523567777$$

The following procedure creates two catenaries through the points $(x0, y0)$ and $(-x0, y0)$. Note that first the correct c must be obtained by solving the equation $y0 = c \cosh(x0/c)$ in its arccos form as above. Note that we can specify an interval (e.g. 0..0.5524341245) in which 'fsolve' looks for a solution. Of course, the two catenaries have u's in different intervals:

```
>  catenary:=proc(x0,y0)
local A,B,c1,c2,p1,p2;
A:=fsolve(x0/y0=u*arccosh(1/u),u,0..0.5524341245);
B:=fsolve(x0/y0=u*arccosh(1/u),u,0.5524341245..1);
c1:=A*y0;
c2:=B*y0;
p1:=plot(c1*cosh(x/c1),x=-x0..x0):
p2:=plot(c2*cosh(x/c2),x=-x0..x0):
display({p1,p2});
end:
```

Now let's see what happens as $x/y \rightarrow .6627434192$:

> catenary(0.3,1);

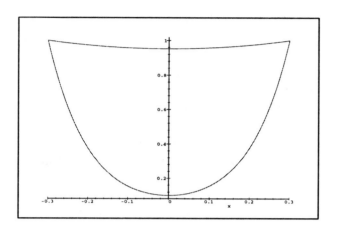

> catenary(0.6,1);

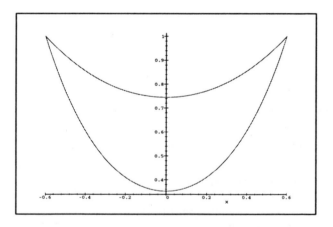

```
> catenary(0.6627,1);
```

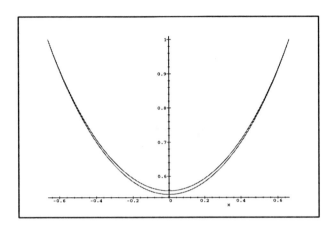

5.6.2. Catenoids. Now let's look at two catenoids obtained from two catenaries having the same value of x/y:

```
> plot3d([.1003411682*cosh(u/.1003411682)*cos(v ),
.1003411682*cosh(u/.1003411682)*sin(v),u],u=-.3..0.3,
v=0..2*Pi,scaling=constrained,style=patch,
orientation=[39,81]);
```

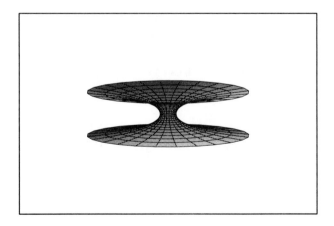

```
>   plot3d([.9523567777*cosh(u/.9523567777)* cos(v),
.9523567777*cosh(u/.9523567777)*sin(v),u],u=-.3..0.3,
v=0..2*Pi,scaling=constrained,style=patch,
orientation=[39,81]);
```

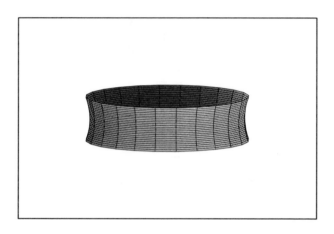

We now show that the narrow waist catenoids (with $u < .5524341245$) always have larger area than the thicker waist catenoids (which grow slower). By the First Principle of Soap Films, this means that the soap films we see are only the thick-waisted ones! First let's compute surface area for catenoids using the usual surface of revolution formula

$$\text{Area} = 2\pi \int y\sqrt{1 + y'^2}\, dx.$$

Again, to non-dimensionalize things, we divide the surface area of the catenoid by that of the two disks centered at $(-x_0, y_0)$ and (x_0, y_0); namely, $2\pi y_0^2$. Also note that we have $\sqrt{1 + y'^2} = \cosh(x/c)$ for $y = c\cosh(x/c)$:

```
>   area_vs_disks=Int(simplify((1/y0)^2*c*cosh(x/ c)*
sqrt(1+diff(c*cosh(x/c),x)^2),symbolic),x=-x0..x0);
```

$$area\_vs\_disks = \int_{-x0}^{x0} \frac{c\cosh(\frac{x}{c})^2}{y0^2}\, dx$$

As examples, let's use the values of c obtained for $x/y = 0.3$ above:

```
>  evalf(Int(.1003411682*cosh(x/.1003411682)^2,
x=-0.3..0.3));
```
$$1.025055441$$

```
>  evalf(Int(.9523567777*cosh(x/.9523567777)^2,
x=-0.3..0.3));
```
$$.5906932099$$

The thick-waisted catenoid has smaller area. In order to see how this works in general, let's parametrically plot x/y versus non-dimensionalized surface area on the narrow waist interval $0..0.5524341245$ and on the thick waist interval $0.5524341245..1$. We also will include a horizontal line (i.e. the Goldschmidt solution) at 1 to represent the area of the two disks. (Remember, this is non-dimensionalized, so we are looking at catenoid area divided by disk area.) Finally, we will include a vertical line at $x/y = 0.3$ so that the values computed above can be checked on the graph. Before we can do this, however, we need a formula for non-dimensionalized catenoid surface area which depends only on $u = c/y$ (as $x/y = (c/y)\mathrm{arccosh}(y/c)$ did). From the formula above, we obtain

$$
\begin{aligned}
\frac{\text{Area}}{2\pi\, y0^2} &= \int_{-x0}^{x0} \frac{c\cosh(x/c)^2}{y0^2}\, dx \\
&= \frac{c}{y0^2} \int_{-x0}^{x0} \frac{1 + \cosh(2x/c)}{2}\, dx \\
&= \frac{c}{2y0^2}\bigg|_{-x0}^{x0}\; x + \frac{c}{2}\sinh(2x/c) \\
&= \frac{c}{2y0^2}\left(2x0 + 2c\,\sinh(x0/c)\cosh(x0/c)\right) \\
&= \frac{c}{y0}\left(\frac{x0}{y0} + \frac{c}{y0}\,\sinh(x0/c)\cosh(x0/c)\right) \\
&= \left(\frac{c}{y0}\right)^2\left(\mathrm{arccosh}(y0/c) + \frac{y0}{c}\sqrt{\left(\frac{y0}{c}\right)^2 - 1}\right) \\
&= u^2\left(\mathrm{arccosh}(1/u) + \frac{1}{u}\sqrt{1/u^2 - 1}\right).
\end{aligned}
$$

```
>  p1:=plot([u*arccosh(1/u),u^2*(arccosh(1/u)+
1/u*sqrt(1/u^2-1)), u=0..0.5524341245],color=red,
style=point):
```

```
> p2:=plot([u*arccosh(1/u),u^2*(arccosh(1/u)+
1/u*sqrt(1/u^2-1)), u=.5524341245..1],color=blue):
> p3:=plot([u,1,u=0..0.68],color=magenta):
> p4:=plot([0.3,u,u=0..1.2],color=green):
> display(p1,p2,p3,p4);
```

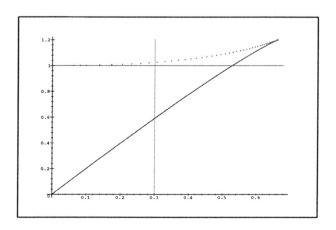

The narrow waist catenoids have areas plotted with a dotted line. So we see now that the narrow waist catenoids always have areas larger than the thick waisted catenoids. In fact, the narrow catenoids have larger areas than the two disks! The thick catenoids have areas smaller than the two disks up to a certain point. What do you think this is? We can also use Maple's 'fsolve' command to find the intersection of the Goldschmidt line and the thick-waisted area line:

```
> c_over_y0:=fsolve(u^2*(arccosh(1/u)+ 1/u*sqrt(1/u^2
-1))=1,u):  evalf(subs(u=c_over_y0,u*arccosh(1/u)));
```

.5276973968

And now we understand the exercise completely — or at least as far as we believe Maple. Of course, analysis can rigorously justify these Maple calculations, but the results above are convincing in themselves.

5.6.3. Another Comparison. Let's look at the area question another way to verify the work above. The surface area of the catenoid obtained by revolving $y = c\cosh(x/c)$ about the x-axis is given by

$$\int_0^{2\pi} \int_{-x0}^{x0} |catenoid\_u \times catenoid\_v| \, du \, dv,$$

where 'catenoid' is the parametrization of the catenoid given below.

```
> catenoid:=[u,a*cosh(u/a)*cos(v), a*cosh(u/a)*
sin(v)];
```

$$catenoid := [u, \, a\cosh(\tfrac{u}{a})\cos(v), \, a\cosh(\tfrac{u}{a})\sin(v)]$$

```
> catenoid_u:=diff(catenoid,u);catenoid_v:=
diff(catenoid,v);
```

$$catenoid\_u := [1, \, \sinh(\tfrac{u}{a})\cos(v), \, \sinh(\tfrac{u}{a})\sin(v)]$$

$$catenoid\_v := [0, \, -a\cosh(\tfrac{u}{a})\sin(v), \, a\cosh(\tfrac{u}{a})\cos(v)]$$

```
> simplify(sqrt(crossprod(catenoid_u, catenoid_v)
[1]^2+ crossprod(catenoid_u,catenoid_v)[2]^2+
crossprod(catenoid_u, catenoid_v)[3]^2),symbolic);
```

$$a\cosh(\tfrac{u}{a})^2$$

So the following calculation gives the surface area of the catenoid:

```
> simplify(expand(convert(int(int(a*cosh(u/a)^2,
u=-x0..x0), v=0..2*Pi),trig)));
```

$$2\,\pi\,a^2\sinh(\tfrac{x0}{a})\cosh(\tfrac{x0}{a}) + 2\,\pi\,a\,x0$$

The following procedure plots the difference in areas of the catenoids and disks obtained when y0=1 and x0 goes from 0 to .662. Larger N produces a finer grid in [0,.662] and we only print a few of the calculations:

```
> area_cat:=proc(N)
local A,areadiff,L,i;
L:=[ ];
for i from 1 to N do
A:=fsolve(1=a*cosh(.662*i/(N*a)),a);
areadiff:=evalf(2*Pi*A^2*sinh(.662*i/(N*A))*
```

```
cosh(.662*i/(N*A))+2*Pi*A*.662*i/N-2*Pi);
L:=[op(L),.662*i/N,areadiff];
if .662*i/N>.45 then print('x0=',.662*i/N,
'areadiff=',areadiff) fi;
od:
pointplot(L,color=red);
end:
> area_cat(50);
```

$$x0 =, .4501600000,\ areadiff =, -.8338639837$$

$$x0 =, .4634000000,\ areadiff =, -.6876941871$$

$$x0 =, .4766400000,\ areadiff =, -.5429646267$$

$$x0 =, .4898800000,\ areadiff =, -.3997604204$$

$$x0 =, .5031200000,\ areadiff =, -.2581747859$$

$$x0 =, .5163600000,\ areadiff =, -.1183106917$$

$$x0 =, .5296000000,\ areadiff =, .01971699552$$

$$x0 =, .5428400000,\ areadiff =, .1557786159$$

$$x0 =, .5560800000,\ areadiff =, .2897257745$$

$$x0 =, .5693200000,\ areadiff =, .4213856733$$

$$x0 =, .5825600000,\ areadiff =, .5505527664$$

$$x0 =, .5958000000,\ areadiff =, .6769756824$$

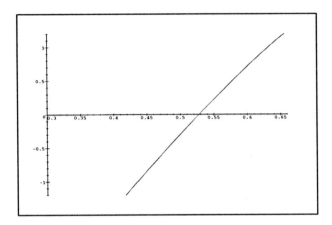

Notice where the negatives turn to positives. In between (i.e. at 0) must be the place where the catenoid and disks have equal area. The plot shows about where this happens — our old friend .5276973968. This is verified by

```
> fsolve(1=a*cosh(x0/a),2*Pi*a^2*sinh(1/a*x0)
*cosh(1/a*x0)+2*Pi*a*x0-2*Pi=0,a,x0);
```

$$a = .8255174537, x0 = .5276973970$$

5.7. Some Minimal Surfaces

In this section we plot some famous minimal surfaces from Example 3.2.10. As usual, we begin with

```
>  with(plots):
```

Let's plot Scherk's *first* surface *first*. Recall from Example 3.2.4 that Scherk's surface is only defined on a checkerboard of squares in the plane. This is true because the logarithm must have positive inputs. We will save the plots on different squares by giving them names and displaying them at one time:

```
>  scherk1:=[u,v,ln(cos(v)/cos(u))]:
>  ps1:=plot3d(scherk1,u=-Pi/2+.001..Pi/2-.001,
v=-Pi/2+.001..Pi/2-.001,grid=[10,10]):
```

Notice that we have to be careful around the singularities — the lines of reflection at the corners of squares where the surface is defined. Now let's save the plots on other squares:

```
>  ps2:=plot3d(scherk1,u=Pi/2+.001..3*Pi/2-.001,
v=Pi/2+.001..3*Pi/2-.001,grid=[10,10]):
>  ps3:=plot3d(scherk1,u=-3*Pi/2+.001..  -Pi/2-.001,
v=-3*Pi/2+.001..-Pi/2-.001,grid=[10,10]):
>  ps4:=plot3d(scherk1,u=Pi/2+.001..3*Pi/2-.001,
v=-3*Pi/2+.001..-Pi/2-.001,grid=[10,10]):
>  ps5:=plot3d(scherk1,u=-3*Pi/2+.001..  -Pi/2-.001,
v=Pi/2+.001..3*Pi/2-.001,grid=[10,10]):
```

Finally, we can display all the plots:

```
>  display({ps1,ps2,ps3,ps4,ps5},style=patch,
orientation=[132,41],scaling=unconstrained,
shading=xy,lightmodel=light2);
```

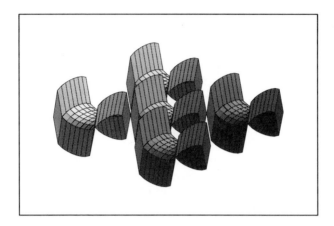

To see the checkerboard, just change the orientation of the plot:

```
>   display({ps1,ps2,ps3,ps4,ps5},style=patch,
orientation=[0,0],scaling=unconstrained,
shading=xy,lightmodel=light2);
```

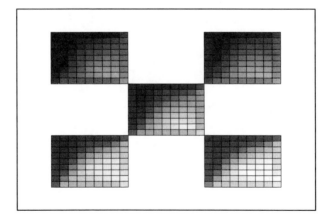

The minimal surfaces of the 1800's, the helicoid and the catenoid, are two faces of the same coin. By this, we mean that there is a continuous deformation of the helicoid into the catenoid which, at each stage, preserves the essential geometry of the surface. Technically speaking, at each stage of the deformation, the resulting surface is isometric to the first stage (but only locally so at the very last stage). See most texts on differential geometry for a broader discussion. To see the deformation (which arises from the Weierstrass-Enneper representation), let's define a procedure:

```
> helcatplot := proc(t)
local X;
X :=[cos(t)*sinh(v)*sin(u)+sin(t)*cosh(v)*cos(u),
-cos(t)*sinh(v)*cos(u)+sin(t)*cosh(v)*sin(u),
u*cos(t)+v*sin(t)];
plot3d(X,u=0..4*Pi,v=-2..2,style=patch, scaling=
constrained, shading=xyz,grid=[40,7],
orientation=[45,54],lightmodel=light3);
end:

> helcatplot(0);
```

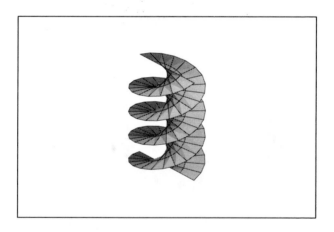

> `helcatplot(3*Pi/14);`

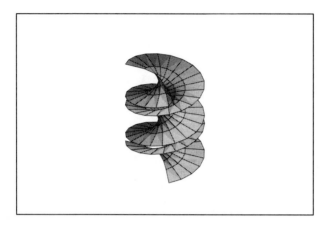

> `helcatplot(2*Pi/7);`

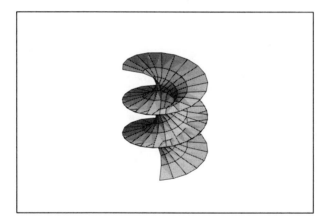

```
> helcatplot(5*Pi/14);
```

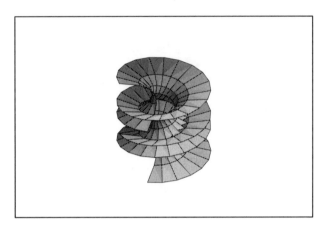

```
> helcatplot(7*Pi/14);
```

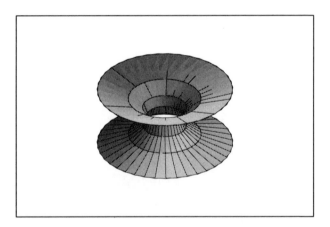

Maple allows an animation of the deformation by using the following command. Note that 'seq' makes a sequence of plots using the procedure 'helcatplot' and then gives an animation by designating 'insequence=true'. For those completely unfamiliar with Maple, when the picture appears on the screen, click on it and an animation toolbar will appear. Then click on the forward arrow:

```
>  display3d(seq(helcatplot(t*Pi/20),t=0..10),
insequence = true);
```

Here is Henneberg's surface:

```
>  henneberg:=[-1+cosh(2*u)*cos(2*v), sinh(u)*sin(v)+
1/3*sinh(3*u)*sin(3*v),-sinh(u)*cos(v)+
1/3*sinh(3*u)*cos(3*v)]:
```

```
>  plot3d(henneberg,u=-1..1,v=0..Pi, grid=[25,30],
orientation=[-59,74],scaling=constrained,style=
patch,shading=xyz, lightmodel=light1);
```

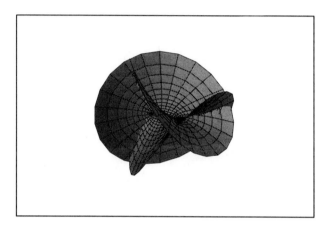

Exercise 5.7.1. Plot Enneper's surface.

Catalan's surface is next:

```
>  catsurf:=[u-sin(u)*cosh(v),1-cos(u)*cosh(v),
4*sin(u/2)*sinh(v/2)]:
```

```
>  plot3d(catsurf,u=0..2*Pi,v=-1.5..1.5, scaling=
constrained,grid=[50,10],shading=xyz,style=patch,
orientation=[-15,60],lightmodel=light2);
```

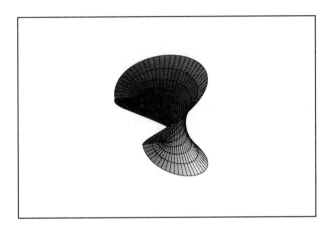

Finally, here is Scherk's fifth surface:

```
>  plot3d([arcsinh(u),arcsinh(v),arcsin(u*v)],
u=-1..1, v=-1..1,scaling=constrained,grid=[12,12],
style=patch,orientation=[72,62],shading=zhue,
lightmodel=light2);
```

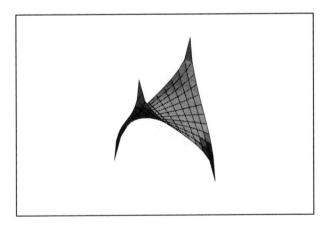

5.8. Enneper's Surface

5.8.1. Enneper's surface. This section shows how Maple may be used to understand the fact that Enneper's surface, within a certain radius boundary curve, is minimal, but not area-minimizing:

```
> with(plots):with(linalg):
> enneper:=[u-u^3/3+u*v^2,-v+v^3/3-v*u^2, u^2-v^2];
```

$$enneper := [u - \tfrac{1}{3}\,u^3 + u\,v^2, \; -v + \tfrac{1}{3}\,v^3 - v\,u^2, \; u^2 - v^2]$$

```
> ennpolar:=simplify(subs({u=r*cos(theta), v=
r*sin(theta)}, enneper));
```

$$ennpolar := [r\cos(\theta) - \tfrac{4}{3}\,r^3\cos(\theta)^3 + r^3\cos(\theta), \; -r\sin(\theta)$$
$$+ \tfrac{1}{3}\,r^3\sin(\theta) - \tfrac{4}{3}\,r^3\sin(\theta)\cos(\theta)^2, 2\,r^2\cos(\theta)^2 - r^2]$$

```
> plot3d(ennpolar,r=0..2.3,theta=0..2*Pi, scaling=
constrained, style=patch,shading=zhue,grid=[10,60],
orientation=[90,70]);
```

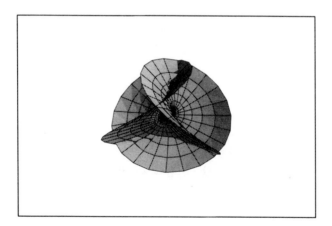

For $r < \sqrt{3}$ it can be shown that enneper does not intersect itself. To see this, consider the expression $x^2 + y^2 + \frac{4}{3}\,z^2$, where x, y, z are the components above, and carry out the following commands:

```
> factor(simplify((r*cos(theta)-4/3*r^3* cos(theta)^3
+r^3*cos(theta))^2+(-r*sin(theta)+1/3*r^3*sin(theta)-
```

```
4/3*r^3*sin(theta)*cos(theta)^2)^2 +4/3*(2*r^2*
cos(theta)^2-r^2)^2));
```

Consider the derivative of this expression with respect to r:

```
> factor(diff(1/9*r^2*(3+r^2)^2,r));
```

For $0 < r$ this is positive, so the function is strictly increasing and, therefore, points with different r values can never give the same point under ennpolar. In fact, it can also be shown that, for $r < \sqrt{3}$, points with the same r cannot map to the same point in 3-space. For points with the same r and different θ_1, θ_2 (i.e. lying on the same circle), look at the third coordinate (in polar form) first and show that $\cos(2\theta_1) = \cos(2\theta_2)$. Then write the first and second coordinates as

$$r \cos\theta \left(1 + r^2 - \frac{4}{3}r^2 \cos^2\theta\right), \quad r \sin\theta \left(1 + r^2 - \frac{4}{3}r^2 \sin^2\theta\right)$$

to see that this would then imply that $\cos\theta_1 = \cos\theta_2$ and $\sin\theta_1 = \sin\theta_2$. Hence, $\theta_1 = \theta_2$. However, note the following:

```
> simplify(subs({r=sqrt(3),theta=0},ennpolar) );
simplify(subs({r=sqrt(3),theta=Pi},ennpolar));
```

$$[0,\ 0,\ 3]$$

$$[0,\ 0,\ 3]$$

So, on the circle with $r = \sqrt{3}$, the points with $\theta = 0$ and $\theta = \pi$ map to the same point in 3-space.

The following represents the Jordan curve which is the boundary curve for both Enneper's surface with $R = 1.5$ and the cylinder defined below. The form of the output differs from MapleV version 4 to Maple 6, but it is the same. Also, note that we take $r = 1.51$ for the input to Maple because this makes the curve easier to see when we display it with the surface.

```
> jorcurve:=subs(r=1.51,ennpolar);
```

$jorcurve := [4.952951 \cos(\theta) - 4.590601332 \cos(\theta)^3,$
$\qquad -.362349667 \sin(\theta) - 4.590601332 \sin(\theta) \cos(\theta)^2, 4.5602 \cos(\theta)^2$
$\qquad -2.2801]$

```
>   bound:=spacecurve(jorcurve,theta=0..2*Pi,
color=black, thickness=3):
>   enn:=plot3d(ennpolar,r=0..1.5,theta=0..2*Pi,
scaling=constrained,grid=[5,50],style=patch):
```

The following procedure creates a parametrization for the cylinder spanning the boundary curve of Enneper's surface with $R = 1.5$. We will write the parametrization explicitly in § 5.8.2, and use it for a calculation of surface area. In order to graph the cylinder, we need two cases — when yv is positive and when yv is negative. The second coordinate yv is negative when v is between 0 and π, so here u must vary from 0 to 2. When v goes from π to 2π, u must vary between -2 and 0:

```
>   CylEnn := proc(r)
local xtheta,ytheta,ztheta,n,X;
xtheta :=r*cos(theta)-1/3*r^3*cos(3*theta);
ytheta :=-r*sin(theta)-1/3*r^3*sin(3*theta);
ztheta := r^2*cos(2*theta);
n := abs(ytheta);
X := [xtheta,ytheta+u*n,ztheta];
end:
>   cyl1:=plot3d(CylEnn(1.5),u=0..2,theta=0..Pi,
scaling=constrained,grid=[5,50],style=patch):
>   cyl2:=plot3d(CylEnn(1.5),u=-2..0, theta=Pi..2*Pi,
scaling=constrained,grid=[5,50],style=patch):
```

Now we can display both surfaces with the same boundary curve:

```
>   display({bound,enn},scaling=constrained, style=
wireframe,orientation=[154,-106]);
```

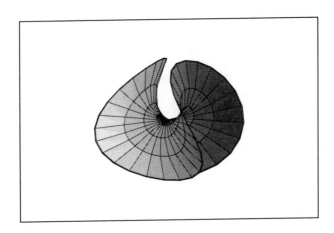

```
>  display({bound,cyl1,cyl2},scaling= constrained,
   orientation=[154,-106]);
```

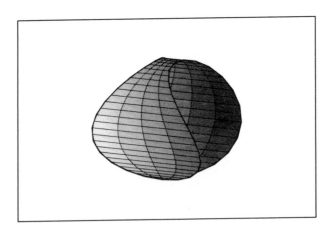

5.8.2. Area Calculation and the Gauss Map. Now let's compute areas. Put $|\mathbf{x}_u \times \mathbf{x}_v| = (1+u^2+v^2)^2$ into polar coordinates with $u = r\cos\theta$ and $v = r\sin\theta$. The surface area of Enneper's surface inside C is given by

$$\text{Area} = \int_0^{2\pi} \int_0^R (r^2+1)^2 r \, dr \, d\theta$$

where the r-factor comes from the Jacobian determinant. To get an explicit number, take $R = 1.5$ say.

Now let's construct another surface with the same boundary curve. In polar coordinates, the curve C has the form

$$C\colon \left(R\cos\theta - \tfrac{1}{3}R^3\cos 3\theta,\ -R\sin\theta - \tfrac{1}{3}R^3\sin 3\theta,\ R^2\cos 2\theta\right).$$

Recall that a general *cylinder* is a surface constructed by taking a base curve (called the *directrix*) and proceeding outwards from each point on the curve in the same direction. A parametrization for a cylinder has the form

$$\mathbf{x}(v, u) = \beta(u) + v\,\mathbf{w},$$

where β is the directrix and $\mathbf{w}$ is a fixed direction vector. The ordinary 'soda can' cylinder is obtained by taking β to be a circle and $\mathbf{w}$ to be a vector perpendicular to the plane in which the circle lies. Because of the symmetry of the boundary curve C above, we can create a cylinder with C as directrix by keeping the x- and z-coordinates fixed and letting the y-coordinate vary between points on C. The second condition says that the fixed direction vector is $\mathbf{w} = (0, 1, 0)$. Therefore, the (non-minimal) cylinder through the boundary curve C has the parametrization

$$y(s, \theta) = \left(R\cos\theta - \tfrac{1}{3}R^3\cos 3\theta,\ -R\sin\theta - \tfrac{1}{3}R^3\sin 3\theta + s,\ R^2\cos 2\theta\right)$$

where $0 \le \theta \le \pi$, $0 \le s \le 2|-R\sin\theta - \tfrac{1}{3}R^3\sin 3\theta|$ and, corresponding to the procedure CylEnn,

$$x(\theta) = R\cos\theta - \tfrac{1}{3}R^3\cos 3\theta,$$
$$y(\theta) = -R\sin\theta - \tfrac{1}{3}R^3\sin 3\theta,$$
$$z(\theta) = R^2\cos 2\theta.$$

The area of the cylinder inside C is given by

$$\text{Area} = \int_0^\pi 2|y(\theta)|\sqrt{x'(\theta)^2 + z'(\theta)^2}\,d\theta.$$

So now we can use Maple to calculate this quantity:

```
>  ytheta:=subs(r=1.5,ennpolar[2]);
```

$$ytheta := -.375000000\sin(\theta) - 4.499999999\sin(\theta)\cos(\theta)^2$$

```
>   x1:=diff(subs(r=1.5,ennpolar[1]),theta);
z1:=diff(subs(r=1.5,ennpolar[3]),theta);
```

$$x1 := -4.875 \sin(\theta) + 13.50000000 \sin(\theta) \cos(\theta)^2$$

$$z1 := -9.00 \cos(\theta) \sin(\theta)$$

```
>   evalf(Int(2*abs(ytheta)*sqrt(x1^2+z1^2),
theta=0..Pi));
```

$$31.66323514$$

On the other hand, Enneper's surface has surface area which may be computed explicitly to be $\pi r^2 (1 + r^2 + \frac{r^4}{3})$ for radius r. Therefore, we have for $r = 1.5$:

```
>   evalf(subs(r=1.5, Pi*r^2*(1+r^2+(r^4)/3)));
```

$$34.90113089$$

This type of comparison can be made more precise to show that Enneper's surface does not minimize area inside the given boundary curve even if we only look at surfaces which remain close to it (in a certain sense). See [**Nit89**] for details. Now here is another viewpoint. Theorem 3.10.6 says that a minimal surface (spanning a curve C) whose Gauss map contains a hemisphere of the unit sphere cannot be area minimizing (among surfaces spanning C). Let's look at the Gauss map of Enneper's surface:

```
>   tanvecs := proc(X)
local Xu,Xv;
Xu :=[diff(X[1],u),diff(X[2],u),diff(X[3],u)];
Xv :=[diff(X[1],v),diff(X[2],v),diff(X[3],v)];
simplify([Xu,Xv]);
end:

>   EFG := proc(X)
local E,F,G,Y;
Y := tanvecs(X);
E := dp(Y[1],Y[1]);
F := dp(Y[1],Y[2]);
G := dp(Y[2],Y[2]);
simplify([E,F,G],sqrt,symbolic);
end:
```

```
>  UN := proc(X)
local Y,Z,s;
Y := tanvecs(X);
Z := crossprod(Y[1],Y[2]);
s := sqrt(Z[1]^2+Z[2]^2+Z[3]^2);
simplify([Z[1]/s,Z[2]/s,Z[3]/s],sqrt,symbolic);
end:
```

First, we find the unit normal (or Gauss map) of Enneper. Then, in order to plot the result out to the boundary curve $r = \sqrt{3}$, we put the result into polar coordinates:

```
>  gaussmap:=simplify(subs({u=r*cos(theta),
v=r*sin(theta)}, UN(enneper)));
```

$$gaussmap := \left[2\,\frac{r\cos(\theta)}{1+r^2},\, 2\,\frac{r\sin(\theta)}{1+r^2},\, \frac{r^2-1}{1+r^2} \right]$$

Now let's compute the derivative of the third coordinate.

```
>  simplify(diff((r^2-1)/(1+r^2),r));
```

$$4\,\frac{r}{(1+r^2)^2}$$

For $r > 0$, the derivative is positive. Therefore, the third coordinate is strictly increasing, and this means that different r's produce different Gauss map images. If, on the other hand, r is fixed and θ varies, then the first two coordinates of the Gauss map tell us that the Gauss map is one-to-one. Now let's graph the Gauss map and then graph it along with the unit sphere. Theorem 3.10.6 then tells us that Enneper's surface cannot have minimum area for all curfaces spanning the boundary curve with $r = \sqrt{3}$.

```
>  plot3d(gaussmap,r=0..sqrt(3),theta=0..2*Pi,
scaling=constrained, style=patch,shading=zhue,
orientation=[33,74]);
```

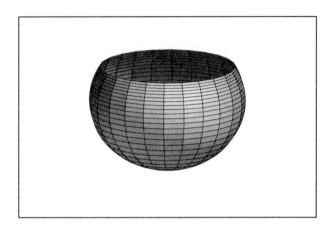

```
>   gm:=plot3d(gaussmap,r=0..sqrt(3), theta=0..2*Pi,
style=patch):
```

```
>   sph:=plot3d([cos(u)*cos(v),sin(u)*cos(v), sin(v)],
u=0..2*Pi,v=-Pi/2..Pi/2,style=patch):
```

```
>   A:=array(1..2):
```

```
>   A[1]:=gm:A[2]:=sph:
```

```
>   display(A,shading=zhue,orientation=[33,74],
scaling=constrained);
```

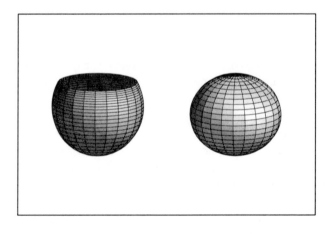

5.9. The Weierstrass-Enneper Representation

5.9.1. The Procedures. This section presents a way to obtain the Weierstrass-Enneper Representations (see Theorem 3.6.6 and Theorem 3.6.7) for minimal surfaces using Maple. As usual, Maple does not always simplify algebra in the most pleasant way, so some algebraic manipulation is usually necessary. The reader may be forgiven for wondering why the sequence of simplifications have been chosen in 'Weier' below. The only answer to be given is that, by trial and error, these steps worked in almost all cases (in MapleV version 4). Having said that, we must point out that these steps work very well for version 4 of MapleV and less well for version 5 and Maple 6 (due to a change in the simplify command, it seems). For some of the situations where this is the case, we will provide alternative version 5 and Maple 6 commands.

```
>  with(plots):with(linalg):
```

We must make the following assumptions to ensure that Maple understands the difference between complex and real variables. The variable z will represent complex numbers while u and v will denote real numbers. We then write $z = u + I\,v$:

```
>  assume(u,real);additionally(v,real); additionally(t,
real);is(u,real);is(v,real);
```

$$true$$

$$true$$

The following procedure simply returns the complex integrals of the WE Representation and may be useful when the simplification process of 'Weier' fails:

```
>  Weierz := proc(F)
local Z1,Z2,Z3,Z;
Z1 := int(F*(1-z^2),z);
Z2 := int(I*F*(1+z^2),z);
Z3 := int(2*F*z,z);
Z  := [Z1,Z2,Z3];
end:
```

The main point of this worksheet is the following procedure, which takes the Weierstrass-Enneper Representation II and attempts to put it in a simple recognizable form. Note that, not only is the complex analytic function $F(z)$ entered as input, but a number 'a' is also inputted. It is often the case that a WE expression may be simplified by changing the variable z to e^z or something like $e^{-i*z/2}$. Below, if $a = 1$, the former is implemented, while if $a = 2$, the latter is implemented. If any other a (such as $a = 0$) is entered, then the procedure simply uses z as z.

```
>  Weier := proc(F,a)
local Z1,Z2,X1,X2,X3,Z3,X;
Z1 := int(F*(1-z^2),z);
Z2 := int(I*F*(1+z^2),z);
Z3 := int(2*F*z,z);
if a=1 then
Z1:=subs(z=exp(z),Z1);
Z2:=subs(z=exp(z),Z2);
Z3:=subs(z=exp(z),Z3) fi;
if a=2 then
Z1:=subs(z=exp(-I*z/2),Z1);
Z2:=subs(z=exp(-I*z/2),Z2);
Z3:=subs(z=exp(-I*z/2),Z3) fi;
X1:=simplify(convert(simplify(Re(evalc(subs(z=u+I*v,
expand(simplify(Z1)))))),trig),trig),trig);
X2:=simplify(convert(simplify(Re(evalc(subs(z=u+I*v,
expand(simplify(Z2)))))),trig),trig),trig);
X3:=simplify(convert(simplify(Re(evalc(subs(z=u+I*v,
expand(simplify(Z3)))))),trig),trig),trig);
X := [X1,X2,X3];
end:
```

Now we can easily write another procedure to handle the Weierstrass-Enneper Representation I :

```
>  Weierfg := proc(f,g,a)
local Z1,Z2,X1,X2,X3,Z3,X;
Z1 := int(f*(1-g^2),z);
Z2 := int(I*f*(1+g^2),z);
Z3 := int(2*f*g,z);
if a=1 then
Z1:=subs(z=exp(z),Z1);
Z2:=subs(z=exp(z),Z2);
```

```
Z3:=subs(z=exp(z),Z3) fi;
if a=2 then
Z1:=subs(z=exp(-I*z/2),Z1);
Z2:=subs(z=exp(-I*z/2),Z2);
Z3:=subs(z=exp(-I*z/2),Z3) fi;
X1:=simplify(convert(simplify(Re(evalc(subs(z=u+I*v,
expand(simplify(Z1)))))),trig),trig),trig);
X2:=simplify(convert(simplify(Re(evalc(subs(z=u+I*v,
expand(simplify(Z2)))))),trig),trig),trig);
X3:=simplify(convert(simplify(Re(evalc(subs(z=u+I*v,
expand(simplify(Z3)))))),trig),trig),trig);
X := [X1,X2,X3];
end:
```

5.9.2. Examples. Identify the following minimal surfaces by the parametrizations obtained from the WE-Rep II. Again, keep in mind that version 4 of MapleV gives the best results. For MapleV 5 and Maple 6, some human intervention may be needed to simplify algebraic expressions — even for something as simple as the second example below. The plotting procedures are invariably affected by these simplification problems, so the reader may have to adjust the simplifications to achieve reasonable pictures.

```
> Weier(1/(2*z^2),1);
```

$$[-\cos(v)\cosh(u),\ -\sin(v)\cosh(u),\ u]$$

```
> Weier(I/(2*z^2),1);
```

$$[\sin(v)\sinh(u),\ -\cos(v)\sinh(u),\ -v]$$

In Maple 6, the helicoid just computed comes out unsimplified in the first coordinate. There seems to be no easy way to remedy this (even using side relations in simplify), but this does not affect plotting. A more serious problem arises in the third coordinate, however. Maple 6 gives $-\arctan(\sin(v),\cos(v))$ (meaning $\arctan(\tan(v))$), which should simplify (at least symbolically) to $-v$ (as it does in version 4). This causes problems in plotting, since arctan has bad behavior near $v = \pi/2$ for instance. The best advice for applying the Weierstrass procedures seems to be this. Compute as we have shown, and whenever an $\arctan(\sin(v),\cos(v))$ comes up, copy the output and replace the arctan with v. This works with the helicoid above, for example.

```
> combine(Weier(1-1/z^4,1),trig);
```

$$[2\cos(v)\sinh(u) - \tfrac{2}{3}\cos(3\,v)\sinh(3\,u),$$
$$- \tfrac{2}{3}\sin(3\,v)\sinh(3\,u) - 2\sin(v)\sinh(u),\ 2\cos(2\,v)\cosh(2\,u)]$$

```
> Weier(1,0);
```

$$[u - \tfrac{1}{3}u^3 + u\,v^2,\ -v - u^2\,v + \tfrac{1}{3}v^3,\ u^2 - v^2]$$

The following example must be treated differently in Maple 6. Namely, instead of using Weier($I/z - I/z^3, 2$), in Maple 6, it is better to use Weier($I/z - I/z^3, 1$) and replace arctan(sin(v), cos(v)) with v.

```
> Weier(I/z-I/z^3,2);
```

$$[-\sin(u)\cosh(v) + u,\ -\cos(u)\cosh(v),\ 4\sin(\tfrac{1}{2}u)\sinh(\tfrac{1}{2}v)]$$

```
> combine(Weier(1/(1-z^4),0));
```

$$\left[\frac{1}{2}\arctan(u,\ 1-v) - \frac{1}{2}\arctan(-u,\ v+1),\ -\frac{1}{2}\arctan(v,\ u+1)\right.$$
$$\left. + \frac{1}{2}\arctan(-v,\ 1-u), \ln\left(\frac{((1+u^2-v^2)^2 + 4\,u^2\,v^2)^{1/4}}{((u-1)^2+v^2)^{1/4}\,((u+1)^2+v^2)^{1/4}}\right)\right]$$

The following is an example of an associated family of minimal surfaces. Note that all that is needed is to multiply the analytic function $F(z)$ by $e^{i\,t}$ for a real parameter t going between two limits.

```
> Weier(exp(I*t)*1/(2*z^2),1);
```

$$[-\cos(t)\cos(v)\cosh(u) + \sin(t)\sin(v)\sinh(u),\ -\sin(t)\cos(v)\sinh(u)$$
$$-\cos(t)\sin(v)\cosh(u),\ \cos(t)\,u - \sin(t)\,v]$$

The associated family can be graphed as the parameter changes by animating the sequence of pictures. Note that the parameter t has been replaced by $\frac{\pi}{2}\frac{i}{10}$ in order to get a sequence of pictures as i changes from 0 to 10.

```
> display([seq(plot3d([ -cos(Pi/2*i/10)*cos(v)*
cosh(u)+sin(Pi/2*i/10)*sin(v)*sinh(u), -sin(v)*
cos(Pi/2*i/10)*cosh(u)-cos(v)*sin(Pi/2*i/10)*sinh(u),
cos(Pi/2*i/10)*u-sin(Pi/2*i/10)*v],u=-1..1,v=0..2*Pi),
i=0..10)],insequence=true);
```

Exercise 5.9.1. Use Weier to create minimal surfaces with $F(z)=$
(1.) z (2.) $1/z$ (3.) $\ln(z)$ (4.) iz (5.) z^2 (6.) $\sin(z)$ (7.) $1/z^3$ (8.)
i/z^3 . Graph these surfaces (choose u and v and rotate and scale to
make your picture as beautiful as possible). As an example:

```
>  Weier(I/z^3,2);
```

$$[-\tfrac{1}{2}u + \tfrac{1}{2}\sin(u)\cosh(v) - \tfrac{1}{2}\sin(u)\sinh(v), \; -\tfrac{1}{2}v + \tfrac{1}{2}\cos(u)\cosh(v)$$
$$-\tfrac{1}{2}\cos(u)\sinh(v), 2\sin(\tfrac{1}{2}u)\cosh(\tfrac{1}{2}v) - 2\sin(\tfrac{1}{2}u)\sinh(\tfrac{1}{2}v)]$$

```
>  plot3d(Weier(I/z^3,2),u=0..6*Pi,v=-1..1, grid=
[80,15],scaling=constrained,orientation=[50,79],
shading=XYZ,style=patch);
```

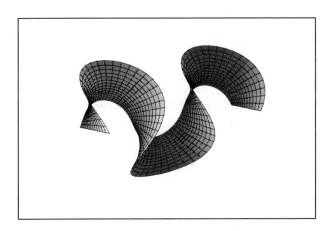

This example is yet another case where option 1 should be used in
place of 2 (i.e. Weier($I/z^3, 1$)) and arctan's should be replaced. Further, once these things are done, it is also helpful to change the limits
on u and v to get the best picture. The reader is invited to experiment.

Exercise 5.9.2. Identify the minimal surfaces below and plot them
using Weier and Weierfg:

```
>  Weierfg(1,z,0);
```

$$[u - \tfrac{1}{3}u^3 + u\,v^2, \; -v - u^2\,v + \tfrac{1}{3}v^3, \; u^2 - v^2]$$

```
>  Weierfg(-exp(-z),-exp(z),0);
```

$$[2\cos(v)\cosh(u),\ 2\sin(v)\cosh(u),\ 2\,u]$$

```
>  Weierfg(1,1/z,1);
```

$$[2\cos(v)\cosh(u),\ -2\sin(v)\cosh(u),\ 2\,u]$$

```
>  Weierfg(-I/2,1/z,1);
```

$$[\sin(v)\sinh(u),\ \cos(v)\sinh(u),\ v]$$

```
>  Weierfg(-I/2*exp(-z),exp(z),0);
```

$$[-\sin(v)\sinh(u),\ \cos(v)\sinh(u),\ v]$$

```
>  Weierfg(1,I/z,1);
```

$$[2\cos(v)\sinh(u),\ -2\sin(v)\sinh(u),\ -2\,v]$$

```
>  simplify(Weierfg(1/z,z,2),trig);
```

$$[-\tfrac{1}{2}\cos(u)\sinh(v)-\tfrac{1}{2}\cos(u)\cosh(v)+\tfrac{1}{2}\,v,$$
$$\tfrac{1}{2}\sin(u)\sinh(v)+\tfrac{1}{2}\sin(u)\cosh(v)-\arctan(-\sin(\tfrac{1}{2}\,u),\,\cos(\tfrac{1}{2}\,u)),$$
$$2\cos(\tfrac{1}{2}\,u)\cosh(\tfrac{1}{2}\,v)+2\cos(\tfrac{1}{2}\,u)\sinh(\tfrac{1}{2}\,v)]$$

Exercise 5.9.3. Plot the surface Weierfg($1/z, z, 2$) with u going from 0 to 4π and v going from -3 to 3.

The next surface is called the *trinoid*. How do you think it got its name? Here is another example of a problem in Maple 6. This time the commands below will provide a plot of the trinoid almost exactly like the version 4 plot, but the time required for plotting is much greater in Maple 6 — on the order of 15 times as long.

```
>  tri:=Weierfg(1/(z^3-1)^2,z^2,0);
>  trinoid:=subs({u=u*cos(v),v=u*sin(v)},tri):
>  plot3d(trinoid,u=0..3.5,v=0..2*Pi,view=[-1..1,
-1..1,-1..1],grid=[25,55],style=patch,shading=zhue,
orientation=[62,64]);
```

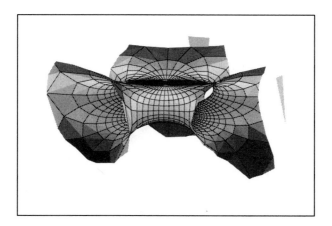

Now let's plot *Richmond's surface*. Note that, in order to get a somewhat decent plot, we need to graph several separate pieces and then display them all at once.

```
> Weierfg(z^2,1/z^2,0);
```

$$\left[\frac{1}{3}\,u^3 - u\,v^2 + \frac{u}{u^2+v^2},\ -u^2\,v + \frac{1}{3}\,v^3 - \frac{v}{u^2+v^2},\ 2\,u\right]$$

```
> rich1:=plot3d(Weierfg(z^2,1/z^2,0),u=-0.5..0.5,
v=.1..1,grid=[30,30]):
> rich2:=plot3d(Weierfg(z^2,1/z^2,0),u=-0.5..0.5,
v=-1..-0.1,grid=[30,30]):
> rich3:=plot3d(Weierfg(z^2,1/z^2,0),u=-0.5..-0.1,
v=-1..1,grid=[40,50]):
> rich4:=plot3d(Weierfg(z^2,1/z^2,0),u=0.1..0.5,
v=-1..1,grid=[40,50]):
> display({rich1,rich2,rich3,rich4},style=
patchnogrid, orientation=[138,83]);
```

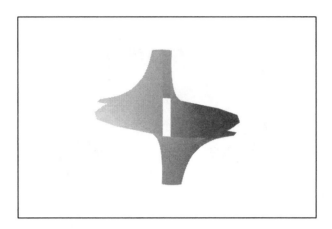

Exercise 5.9.4. Now try the following to see Richmond's surface from another perspective.

```
> Weierfg(z^2,1/z^2,1);
> plot3d(Weierfg(z^2,1/z^2,1),u=-1..1, v=0..2*Pi,
grid=[25,45]);
> Weierfg(z^2,1/z^2,2);
> plot3d(Weierfg(z^2,1/z^2,2),u=0..2*Pi, v=-2..2,
grid=[45,25]);
> Weierfg(1/(1-z^4),z,0);
> hennass:=combine(simplify(Weier(-I*(1-1/z^4), 1),
trig),trig);
> plot3d(hennass,u=-1..1,v=0..Pi,grid=[25,50]);
```

Exercise 5.9.5. Here is a *higher order Enneper's surface*. Plot it and compare to Enneper's surface.

```
> Weierfg(z^2,z^2,0);
```

Here are some extra representations and plots. The first is Catalan's surface, which the observant reader recognized above. Recall that instead of using Weier($I/z - I/z^3, 2$), in Maple 6, it is better to use Weier($I/z - I/z^3, 1$) and replace arctan(sin(v), cos(v)) with v.

The second surface is Scherk's first surface , but note that the graph
obtained from the given representation is somewhat unsatisfactory.

```
> plot3d(Weier(I*(1/z-1/z^3),2),v=-2.1..2.1, u=
0..4*Pi,grid=[15,60],scaling=constrained,style=patch,
shading=xy,lightmodel=light2,orientation=[-78,54]);
```

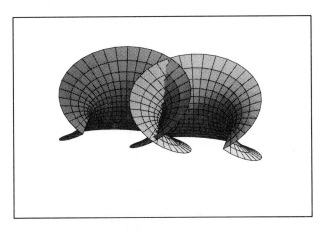

```
> plot3d(Weierfg(4/(1-z^4),z,1),u=-3..1,v=0..2*Pi,
grid=[20,40],scaling=constrained,style=patch,
shading=xy,lightmodel=light2,orientation=[-74,61]);
```

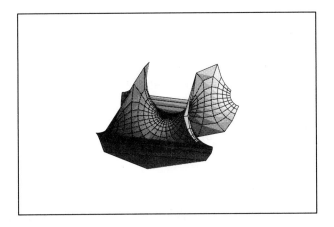

Now, let's make up any function which comes to mind. Notice that, by seeing the parametrization before graphing, we are able to choose the bounds on u and v appropriately.

```
>   Weier(I*(1/z+1/z^2-1/z^3),2);
```

$$[-\sin(u)\cosh(v) - 2\sin(\tfrac{1}{2}u)\sinh(\tfrac{1}{2}v) + u, \ -\cos(u)\cosh(v)$$
$$-2\cos(\tfrac{1}{2}u)\sinh(\tfrac{1}{2}v), 4\sin(\tfrac{1}{2}u)\sinh(\tfrac{1}{2}v) + u]$$

```
>   plot3d(Weier(I*(1/z+1/z^2-1/z^3),2),u=0..4*Pi,
    v=-3..2, grid=[60,15],scaling=constrained,
    style=patch,shading=xy,lightmodel=light2,
    orientation=[-76,86]);
```

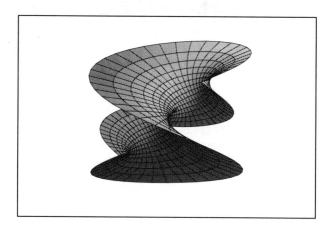

The three examples just given do not work well in version 5 of Maple. As before, the Maple arctan(y, x) function appears where it did not in version 4. The best way to get around this problem is to use the following, for example.

```
>   Weier(I*(1/z-1/z^3),1);
```

$$[4\cos(v)\sin(v)\cosh(u)^2 - 2\cos(v)\sin(v) - 2\arctan(\sin(v), \cos(v)),$$
$$-4\cos(v)^2\cosh(u)^2 + 2\cosh(u)^2 + 2\cos(v)^2 - 1, -4\sin(v)\sinh(u)]$$

```
>   plot3d(Weier(I*(1/z-1/z^3),1),u=-1..1,v=-Pi..Pi,
    grid=[15,60],scaling=constrained,style=patch,shading=
    xy,lightmodel=light2, orientation=[-78,54]);
```

Note that arctan(sin(v), cos(v)) will only be v (in Maple) when $-\pi <$ $v < \pi$, and that is why v is chosen to run from $-\pi$ to π in the plot. Also, arctan(sin(v), cos(v)) may be replaced by v if desired. So for version 5 or Maple 6, the option 1 in 'Weier' is often preferable to 2. The same holds for the two following examples.

```
>  Weierfg(4/(1-z^4),z,1);
```

$$[2\arctan(\cos(v)\cosh(u) + \cos(v)\sinh(u),\, 1 - \sin(v)\cosh(u)$$
$$- \sin(v)\sinh(u)) - 2\arctan(-\cos(v)\cosh(u) - \cos(v)\sinh(u),$$
$$\sin(v)\cosh(u) + \sin(v)\sinh(u) + 1),\, -2\arctan(\sin(v)\cosh(u)$$
$$+ \sin(v)\sinh(u), \cos(v)\cosh(u) + \cos(v)\sinh(u) + 1)$$
$$+ 2\arctan(-\sin(v)\cosh(u) - \sin(v)\sinh(u),\, 1 - \cos(v)\cosh(u)$$
$$- \cos(v)\sinh(u)),\, -\ln(-2\cos(v)\cosh(u) - 2\cos(v)\sinh(u)$$
$$+ 2\cosh(u)^2 + 2\cosh(u)\sinh(u)) - \ln(2\cos(v)\cosh(u)$$
$$+ 2\cos(v)\sinh(u) + 2\cosh(u)^2 + 2\cosh(u)\sinh(u))$$
$$+ \ln(2\cos(2\,v)\cosh(2\,u) + 2\cos(2\,v)\sinh(2\,u) + 2\cosh(2\,u)^2$$
$$+ 2\cosh(2\,u)\sinh(2\,u))]$$

```
>  plot3d(Weierfg(4/(1-z^4),z,1),u=-3..1, v=0..2*Pi,
grid=[20,40],scaling=constrained,style=patch,shading=
xy,lightmodel=light2, orientation=[-74,61]);
>  Weier(I*(1/z+1/z^2-1/z^3),1);
```

$$[4\cos(v)\sin(v)\cosh(u)^2 - 2\arctan(\sin(v),\, \cos(v))$$
$$+2\sin(v)\sinh(u) - 2\cos(v)\sin(v),\, -1 + 2\cos(v)^2$$
$$-4\cos(v)^2\cosh(u)^2 + 2\cosh(u)^2 - 2\cos(v)\sinh(u),$$
$$-4\sin(v)\sinh(u) - 2\arctan(\sin(v),\, \cos(v))]$$

```
>  plot3d(Weier(I*(1/z+1/z^2-1/z^3),1),u=-1..1,
v=-Pi+0.01..Pi-0.01,grid=[10,70],scaling=constrained,
style=patch,shading=xy,lightmodel=light2,orientation=
[-76,86]);
```

Note that here we had to change the range of v a bit to avoid singularities. Now let's return to the general case, always keeping in mind the difference between versions 4 and 5 (as well as Maple 6). Here is another interesting surface.

```
> plot3d(Weierfg(1/z^2,1/z^2,2),u=0..4*Pi, v=-0.5..3,
grid=[60,10],scaling=constrained,style=patch,shading=
xy,lightmodel=light2,orientation=[38,60]);
```

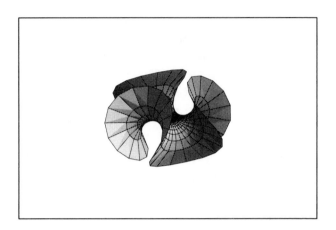

Generalize a bit and see what you get:

```
> plot3d(Weierfg(1/z^2,1/z^3,2),u=0..4*Pi, v=0..8,
grid=[80,15],scaling=constrained,style=patch,shading=
xy,lightmodel=light2,orientation=[-74,61]);
```

The following surface is called *the Bat*:

```
> Weierfg(z,z^3,0);
```

$$[-\tfrac{1}{8}\,u^8 + \tfrac{7}{2}\,u^6\,v^2 - \tfrac{35}{4}\,u^4\,v^4 + \tfrac{7}{2}\,u^2\,v^6 - \tfrac{1}{8}\,v^8 + \tfrac{1}{2}\,u^2 - \tfrac{1}{2}\,v^2,$$
$$-u^7\,v + 7\,u^5\,v^3 - 7\,u^3\,v^5 + u\,v^7 - u\,v,\ \tfrac{2}{5}\,u^5 - 4\,u^3\,v^2 + 2\,u\,v^4]$$

```
> plot3d(Weierfg(z,z^3,0),u=-1..1,v=-1..1,
grid=[40,40],scaling=constrained,style=patch,shading=
xy,lightmodel=light2,orientation=[0,53]);
```

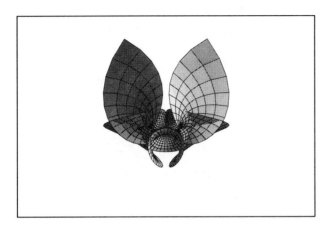

Often, simple analytic functions produce symmetry in the Weierstrass-Enneper representations:

```
>  plot3d(Weier(z^4,0),u=-1..1,v=-1..1,
grid=[40,40],scaling=constrained,style=patch,shading=
xy,lightmodel=light2,orientation=[0,43]);
```

Finally, recall that Henneberg's surface was not orientable because it contained a Moebius strip. The following is a minimal Moebius strip:

```
>  plot3d(Weierfg(I*(z+1)^2/z^4,z^2*(z-1)/(z+1), 1),
u=-0.5..0.5,v=0..Pi,grid=[10,40],scaling=constrained,
style=patch,shading=xy,lightmodel=light2,
orientation=[64,48]);
```

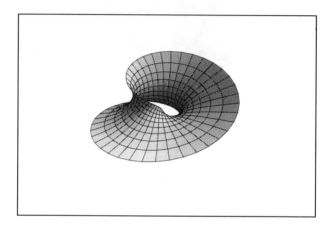

5.10. Björling's Problem

5.10.1. A Procedure for Plane Curves. This section presents a way to obtain the solution to the special case of Björling's problem for plane curves using Maple. As usual, Maple does not always simplify algebra in the most pleasant way, so some algebraic manipulation is usually necessary. Nevertheless, the procedure 'Bjor' below does handle some familiar cases.

```
> with(plots):with(linalg):
```

We must make the following assumptions to ensure that Maple understands the difference between complex and real variables. The variable z will represent complex numbers while u and v will denote real numbers. We then write $z = u + I\,v$.

```
> assume(u,real);additionally(v,real); additionally(t,
  real);is(u,real);is(v,real);
```

Finally, the following procedure takes the solution to Björling's problem for plane curves and their unit principal normal vector fields and attempts to put it in a simple recognizable form. Note that, not only is the parametrized curve $\alpha(t) = [\beta(t),\ \gamma(t),\ 0]$ entered as input, but a number 'a' is also inputted. It is sometimes the case that an expression may be simplified by changing the variable z to e^z or something like $e^{-I*z/2}$. Below, if $a = 1$, the former is implemented, while if $a = 2$, the latter is implemented. If any other a (such as $a = 0$) is entered, then the procedure simply uses z as z.

```
> Bjor := proc(alpha,a)
  local Z1,Z2,X1,X2,X3,Z3,X;
  Z1 := subs(t=z,alpha[1]);
  Z2 := int(sqrt(diff(subs(t=z,alpha[1]),z)^2+
  diff(subs(t=z,alpha[3]),z)^2),z);
  Z3 := subs(t=z,alpha[3]);
  if a=1 then
  Z1:=subs(z=exp(z),Z1);
  Z2:=subs(z=exp(z),Z2);
  Z3:=subs(z=exp(z),Z3) fi;
  if a=2 then
  Z1:=subs(z=exp(-I*z/2),Z1);
  Z2:=subs(z=exp(-I*z/2),Z2);
  Z3:=subs(z=exp(-I*z/2),Z3) fi;
```

```
X1:= simplify(convert(simplify(Re(evalc(subs(z=u+I*v,
expand(simplify(Z1)))))),trig),trig),trig);
X2:=simplify(convert(simplify(Im(evalc(subs(z=u+I*v,
expand(simplify(Z2)))))),trig),trig),trig);
X3:=simplify(convert(simplify(Re(evalc(subs(z=u+I*v,
expand(simplify(Z3)))))),trig),trig),trig);
X := [X1,X2,X3];
end:
```

5.10.2. Examples. The following examples are minimal surfaces
given by the parametrizations obtained from the solution to Björling's
problem. The pictures (when the commands are executed and display
is used) show both the minimal surface and the plane curve generating
it. The first example is the catenoid. Try it.

```
>   cate:=Bjor([cos(t),0,sin(t)],0);
>   cate1:=plot3d(cate,u=0..2*Pi,v=-1..1,grid=[30,10]):
>   cate2:=spacecurve([cos(t),0,sin(t)],t=0..2*Pi,
color=black, thickness=2):
>   display({cate1,cate2});
```

Note that we can use the 'wireframe' style to see both the curve and
surface. Just click on the picture and choose 'style' on the menu bar.
The second example to try is Catalan's surface. Note that the cycloid
is a geodesic in the surface.

```
>   cata:=Bjor([1-cos(t),0,t-sin(t)],0);
>   cata1:=plot3d(cata,u=0..4*Pi,v=-2..2,grid=[50,15]):
>   cata2:=spacecurve([1-cos(t),0,t-sin(t)],t=0..4*Pi,
color=black,thickness=2):
>   display({cata1,cata2});
```

Here is a surface which we will graph generated by the cusp curve.
Compare this to Björling's solution for Neil's parabola (the cusp in
another form). Also, here Maple 6 does not provide as good a picture
as MapleV version 4.

```
>   cuspy:=Bjor([t^2,0,t^3],2);
>   cus1:=plot3d(cuspy,u=0..2*Pi,v=-0.75..0.75, grid=
[50,15]):
```

```
>   cus2:=spacecurve([t^2,0,t^3],t=-1..1,color=black,
thickness=2):
>   display({cus1,cus2});
```

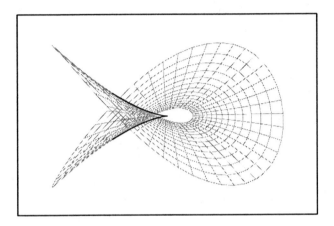

Again note that the 'wireframe' style was used for the picture. Here is a surface generated by the astroid. The commands which generate the picture are listed below it. Compare this to the adjoint surface to Henneberg's surface (Exercise 3.6.24).

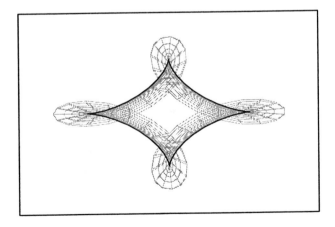

```
>   ast:=Bjor([cos(t)^3,0,sin(t)^3],0);
```

```
>  ast1:=plot3d(ast,u=0..2*Pi,v=-0.5..0.5,grid=
[50,15]):

>  ast2:=spacecurve([cos(t)^3,0,sin(t)^3], t=0..2*Pi,
color=black, thickness=2):

>  display({ast1,ast2},orientation=[90,90],
shading=zhue);
```

The final two examples for the reader to try use a parabola and a spiral.

```
>  para:=Bjor([t,0,t^2],1);

>  para1:=plot3d(para,u=-1..1,v=0..2*Pi,
grid=[15,35]):

>  para2:=spacecurve([t,0,t^2],t=-1..1,
color=black,thickness=2):

>  display({para1,para2});

>  spir:=Bjor([exp(-t)*cos(t),0,exp(-t)*sin(t)], 0);

>  spir1:=plot3d(spir,u=0..4,v=-7..7,grid=[30,55 ]):

>  spir2:=spacecurve([exp(-t/20)*cos(t),0,
exp(-t/20)*sin(t)],t=0..6*Pi, color=black,
thickness=2,numpoints=100):

>  display({spir1,spir2},view=[-1..1,-0.5..0.5,
-1..1],orientation=[95,84]);
```

5.11. The Euler-Lagrange Equations

5.11.1. The Basic Equation. In this section, we show how Maple may be used to obtain and solve the Euler-Lagrange equations. Below, the symbol 'dx' is used for 'x dot' (i.e. $\dot{x}$) and all integrals are of the form

$$J = \int_{t_0}^{t_1} f(t, x(t), \dot{x}(t))\, dt.$$

The reason for this is two-fold. First, the calculus of variations finds use in a much wider mathematical and physical context than what we have discussed. In particular, variational methods are very important in mechanics, and it is typically the case there that the independent variable in the problem is time, t. To make the procedures below as useful as possible to a wide audience, we have decided to keep the mechanics notation. The second reason is that some of the procedures below appeared in [**Opr97**], so it seems best to be consistant with the usage in that text. At any rate, without doing complicated string manipulations in Maple, some fixed variables had to be chosen, so t, x and $\dot{x}$ it is.

The ordinary Euler-Lagrange equation is given by the following procedure. The substitutions '$x = x(t)$' etc. are necessary because the command 'dsolve' used to solve differential equations requires an explicit parameter such as 't'. (This may not be necessary in later versions of Maple.) Also, there is something to note here about the command 'dsolve' in versions 4 and 5 of MapleV. In version 4, Maple is content with giving an implicit solution as output. Usually this is more convenient than taking into account all the possible algebraic manipulations necessary for an explicit solution. Nevertheless, version 5's 'dsolve' has been set up to try to give explicit solutions. This often produces complex solutions which are not relevant to the problem. To get around this difficulty with version 5, try 'dsolve(eq,x(t),implicit)', the extra argument 'implicit' giving an implicit solution. Of course, many times the situation simply doesn't arise, but the reader should be prepared nonetheless.

```
>  EL1:=proc(f)
local part1,part2,p2,dfdx,dfddx;
part1:=subs({x=x(t),dx=diff(x(t),t)},diff(f,x));
```

```
dfdx:=diff(f,dx);
dfddx:=subs({x=x(t),dx=dx(t)},dfdx);
p2:=diff(dfddx,t);
part2:=subs(dx(t)=diff(x(t),t),p2);
RETURN(simplify(part1-part2=0));
end:
```

5.11.2. Some Examples. Now it's time to try some examples.

Example 5.11.1. The first example has integrand $f(t, x, \dot{x})$ given by

```
>  f:=dx^2+2*x*sin(t);
```

$$f := dx^2 + 2\,x\sin(t)$$

with boundary conditions $x(0) = 0$ and $x(\pi) = \pi$.

```
>  eq:=EL1(f);
```

$$eq := 2\sin(t) - 2\left(\frac{\partial^2}{\partial t^2}\,\mathrm{x}(t)\right) = 0$$

Now we can use Maple's 'dsolve' command to solve the differential equation with respect to the initial and final conditions:

```
>  dsolve({eq,x(0)=0,x(Pi)=Pi},x(t));
```

$$\mathrm{x}(t) = -\sin(t) + t$$

Example 5.11.2. Now let's take a more complicated example, but one which we can solve two ways. The first way is to simply use EL1:

```
>  g:=dx*x^2 + dx^2*x;
```

$$g := dx\,x^2 + dx^2\,x$$

```
>  eq2:=EL1(g);
```

$$eq2 := -\left(\frac{\partial}{\partial t}\,\mathrm{x}(t)\right)^2 - 2\left(\frac{\partial^2}{\partial t^2}\,\mathrm{x}(t)\right)\mathrm{x}(t) = 0$$

```
>  dsolve({eq2,x(0)=1,x(1)=4},x(t));
```

Note that no solution comes out. This must be because Maple is confused trying to evaluate and solve for the constants. Instead of putting the endpoint conditions in 'dsolve', let's evaluate them later.

In Maple 6, many confusing solutions come out because 'dsolve' tries to give an explicit answer by default (thus taking roots of roots, etc.). In Maple 6 it is better to use the command 'dsolve(-,implicit)' to obtain the old version 4 answer and then proceed as follows:

```
> dsolve({eq2},x(t));
```

$$t = \frac{2}{3} \frac{\mathrm{x}(t)^{3/2}}{\_C1} - \_C2$$

Now we can substitute in the values of t and x and solve for the arbitrary constants:

```
> solve({subs({t=0,x=1},t = 2/3*x^(3/2)/_C1-_C2),
subs({t=1,x=4},t = 2/3*x^(3/2)/_C1-_C2)},{_C1,_C2});
```

$$\left\{ \_C2 = \frac{1}{7}, \ \_C1 = \frac{14}{3} \right\}$$

It is often useful to isolate a variable in an expression. Maple has an 'isolate' command for that purpose. First we must read it into the Maple session:

```
> readlib(isolate):
> isolate(subs({_C2 = 1/7, _C1 = 14/3}, t =
2/3*x(t)^(3/2)/_C1-_C2),x(t));
```

$$\mathrm{x}(t) = (7\,t + 1)^{2/3}$$

And this is the final extremal for the problem. We will check this in the next subsection by using the first integral approach.

5.11.3. The First Integral. The following procedure handles the case where f does not depend on t:

```
> EL2:=proc(f)
local part1,part2,dfddx ;
part1:=subs({x=x(t),dx=diff(x(t),t)},f);
dfddx:=diff(f,dx);
part2:=diff(x(t),t)*subs({x=x(t),dx=diff(x(t),t)},
dfddx);
RETURN(simplify(part1-part2=c));
end:
```

Example 5.11.3. Let's take the function $g = \dot{x}\,x^2 + \dot{x}^2\,x$ we considered above:

```
>   eq3:=lhs(EL2(g));
```

$$eq3 := -\left(\frac{\partial}{\partial t}\,\mathrm{x}(t)\right)^2 \mathrm{x}(t)$$

```
>   dsolve({eq3=c},x(t));
```

$$\frac{2}{3}\frac{(-\mathrm{x}(t)\,c)^{3/2}}{c^2} + t = \_C1, \qquad -\frac{2}{3}\frac{(-\mathrm{x}(t)\,c)^{3/2}}{c^2} + t = \_C1$$

```
>   solve({subs({t=0,x=1}, -2/3/c^2*(-x*c)^(3/2)+t
=  _C1), subs({t=1,x=4},-2/3/c^2*(-x*c)^(3/2)+t
=  _C1)},{c,_C1});
```

$$\left\{\_C1 = \frac{-1}{7},\, c = \frac{-196}{9}\right\}$$

```
>   simplify(isolate(subs({_C1 = -1/7, c = -196/9},
-2/3/c^2*(-x(t)*c)^(3/2)+t = _C1),x(t)));
```

$$\mathrm{x}(t) = (7\,t + 1)^{2/3}$$

And this verifies our previous calculation.

5.11.4. The Least Area Surface of Revolution. We can now use EL1 and EL2 to discover what a least area surface of revolution is. Recall from Example 4.2.7 and Example 4.4.7 that the surface area of a surface of revolution with profile curve $y(x)$ is given by

$$A = 2\pi \int y\sqrt{1 + y'^2}\,dx.$$

We can neglect the 2π and extremize $\int y\sqrt{1 + y'^2}\,dx$ to find the profile curve y. Remembering our conventions for writing functions to be put into EL1, we write the integrand as

```
>   minrs:=x*sqrt(1+dx^2);
```

$$minrs := x\,\sqrt{1 + dx^2}$$

Now let's apply EL1 to get the Euler-Lagrange equation for the problem:

```
>  meq:=EL1(minrs);
```

$$meq := \frac{1 + (\frac{\partial}{\partial t}\,\mathrm{x}(t))^2 - \mathrm{x}(t)\,(\frac{\partial^2}{\partial t^2}\,\mathrm{x}(t))}{(1 + (\frac{\partial}{\partial t}\,\mathrm{x}(t))^2)^{3/2}} = 0$$

We take the numerator of the expression, set it to zero and solve:

```
>  meq2:=numer(lhs(meq));
```

$$meq2 := 1 + \left(\frac{\partial}{\partial t}\,\mathrm{x}(t)\right)^2 - \mathrm{x}(t)\left(\frac{\partial^2}{\partial t^2}\,\mathrm{x}(t)\right)$$

```
>  dsolve(meq2=0,x(t));
```

$$t = \frac{\ln(\sqrt{\_C1}\,\mathrm{x}(t) + \sqrt{-1 + \mathrm{x}(t)^2\,\_C1})}{\sqrt{\_C1}} - \_C2,$$

$$t = -\frac{\ln(\sqrt{\_C1}\,\mathrm{x}(t) + \sqrt{-1 + \mathrm{x}(t)^2\,\_C1})}{\sqrt{\_C1}} - \_C2$$

Now we solve for $x(t)$ using Maple's 'isolate' command. (Remember that Maple's output is editable, so simply use copy and paste to put previous output into a command as input.) In Maple 6 the answer above is already in exponential notation, so these next few steps are unnecessary. Simply skip to the 'convert' command below if you are using Maple 6.

```
>  readlib(isolate):
>  isolate(t = 1/_C1^(1/2)*ln(_C1^(1/2)*x(t)+
   (-1+x(t)^2*_C1)^(1/2))-_C2,x(t));
```

$$\mathrm{x}(t) = \frac{1}{2}\,\frac{1 + (e^{(\sqrt{\_C1}\,(t+\_C2))})^2}{\sqrt{\_C1}\,e^{(\sqrt{\_C1}\,(t+\_C2))}}$$

Finally, let's convert the exponential expression into one involving hyperbolic trigonometric functions:

```
>  simplify(convert(1/2*(1+exp(_C1^(1/2)*(t+_C2) )^2)/
   _C1^(1/2)/exp(_C1^(1/2)*(t+_C2)),trig));
```

$$\frac{\cosh(\sqrt{\_C1}\,(t + \_C2))}{\sqrt{\_C1}}$$

And this verifies that a least area surface of revolution must be a catenoid (subject to § 5.6). Also see § 5.5.2. Of course, the integrand of the area integral does not depend on the independent variable x explicitly, so we can also use the first integral procedure EL2 to find the profile curve which generates a least area surface:

```
>  meq3:=EL2(minrs);
```

$$meq3 := \frac{\mathrm{x}(t)}{\sqrt{1 + (\frac{\partial}{\partial t}\, \mathrm{x}(t))^2}} = c$$

We know that $x(t) > 0$, so we can make the following assumption and then solve:

```
>  assume(c>0);
>  dsolve(x(t)/(1+diff(x(t),t)^2)^(1/2) = c,x(t));
```

$$\ln(\mathrm{x}(t) + \sqrt{-c^2 + \mathrm{x}(t)^2})\, c + t = \_C1,$$

$$-\ln(\mathrm{x}(t) + \sqrt{-c^2 + \mathrm{x}(t)^2})\, c + t = \_C1$$

```
>  isolate(ln(x(t)+(-c^2+x(t)^2)^(1/2))*c+t =C1,
x(t));
```

$$\mathrm{x}(t) = \frac{1}{2}\, \frac{c^2 + (e^{(-\frac{-\_C1+t}{c})})^2}{e^{(-\frac{-\_C1+t}{c})}}$$

Exercise 5.11.4. Show that this expression is equivalent to $x(t) = c\cosh(t/c + d)$ for some d. Hint: expand cosh into exponential terms and choose d to give the c^2 in the other expression. Once again we see that Maple's simplify command does not always produce output which is friendliest to human eyes.

5.11.5. An Isoperimetric Problem.

Example 5.11.5. This example shows how isoperimetric problems may be treated with Maple. Let's take the function f from our first example above and require that $\int_0^\pi x\, dt = \pi^2 - 2$. The constrained integrand then becomes

```
>  h:=dx^2+2*x*sin(t) - lambda*x;
```

$$h := dx^2 + 2\, x\sin(t) - \lambda\, x$$

```
> eq4:=EL1(h);
```

$$eq4 := 2\sin(t) - \lambda - 2\left(\frac{\partial^2}{\partial t^2}x(t)\right) = 0$$

```
> dsolve({eq4,x(0)=0,x(Pi)=Pi},x(t));
```

$$x(t) = -\sin(t) - \frac{1}{4}\lambda t^2 + \left(\frac{1}{4}\lambda\pi + 1\right)t$$

Now let's use the integral constraint to determine λ:

```
> int(-sin(t)-1/4*lambda*t^2+(1/4*lambda*Pi+1)*t,
t=0..Pi)=Pi^2-2;
```

$$-2 + \frac{1}{24}\lambda\pi^3 + \frac{1}{2}\pi^2 = \pi^2 - 2$$

```
> expand(solve(int(-sin(t)-1/4*lambda*t^2+(1/4*
lambda*Pi+1)*t,t=0..Pi)=Pi^2-2,lambda));
```

$$\frac{12}{\pi}$$

So $\lambda = 12/\pi$, and we can substitute this back into the extremal:

```
> x(t)=subs(lambda=12/Pi,-sin(t)-1/4*lambda*t^2 +
(1/4*lambda*Pi+1)*t);
```

$$x(t) = -\sin(t) - 3\frac{t^2}{\pi} + 4t$$

Example 5.11.6. This example considers the plane curve which minimizes bending energy while keeping a fixed amount of total angular change. In the following, 'dx' represents the derivative of x with respect to arclength t, $\dot{x}$. Also, x denotes the angle made by the tangent to a curve alpha with the horizontal axis. Then 'dx' is known to be the curvature of the curve. The bending energy is the integral of the square of the curvature. That is, we want to minimize this bending energy $\int \frac{\dot{x}^2}{2} dt$ subject to the condition that total angular change $\int x\, dt$ is fixed at some value (say $\frac{1}{6}$). The integrand uses a Lagrange multiplier and is

> `Lag:=1/2* dx^2-lambda*x;`

$$Lag := \frac{1}{2} \, dx^2 - \lambda \, x$$

> `EL1(Lag);`

$$-\lambda - \left(\frac{\partial^2}{\partial t^2} \, \mathrm{x}(t) \right) = 0$$

> `de:=dsolve({EL1(Lag),x(0)=0,x(1)=0},x(t));`

$$de := \mathrm{x}(t) = -\frac{1}{2} \, \lambda \, t^2 + \frac{1}{2} \, \lambda \, t$$

> `In:=int(rhs(de),t=0..1);`

$$In := \frac{1}{12} \, \lambda$$

> `L:=solve(In=1/6,lambda);`

$$L := 2$$

> `xx:=subs(lambda=L,de);`

$$xx := \mathrm{x}(t) = -t^2 + t$$

> `sh:=dsolve({D(x)(s)=cos(-s^2+s), D(y)(s)=`
> `sin(-s^2+s),x(0)=0,y(0)=0},{x(s),y(s)},type=numeric):`

Before the next command can be used, 'with(plots)' must be declared. This command then plots the numeric solution 'sh' of the differential equation above:

> `odeplot(sh, [x(s),y(s)],-5..7,view=[-1..2,-1..1],`
> `numpoints=400);`

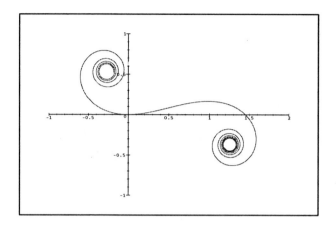

This curve is the *spiral of Cornu*.

Example 5.11.7 (The Catenary). The final example in this section is our old friend the catenary. Let's name the integrand of the integral to be extremized (see Example 4.5.5)

```
>   catenary:=(x-lambda)*sqrt(1+dx^2);
```

$$catenary := (x - \lambda) \sqrt{1 + dx^2}$$

```
>   eq5:=EL2(catenary);
```

$$eq5 := -\frac{-x(t) + \lambda}{\sqrt{1 + (\frac{\partial}{\partial t} x(t))^2}} = c$$

Now let's solve the differential equation. We write it in a form easier for Maple to handle:

```
>   dsolve(x(t)-lambda=c*sqrt(1+diff(x(t),t)^2), x(t));
```

$$-\text{arcsinh}(\frac{\sqrt{-c^2 + x(t)^2 - 2\lambda x(t) + \lambda^2}}{c}) c + t = \_C1,$$

$$-\text{arcsinh}(-\frac{\sqrt{-c^2 + x(t)^2 - 2\lambda x(t) + \lambda^2}}{c}) c + t = \_C1$$

Let's isolate the $x(t)$:

```
>   expand(isolate(-arcsinh(1/c*(-c^2+x(t)^2-
2*lambda*x(t)+lambda^2)^(1/2))*c+t=_C1,x(t)));
```

$$x(t) = \frac{1}{2}\frac{e^{(\frac{-C1}{c})}c}{e^{(\frac{t}{c})}} + \frac{1}{2}\frac{e^{(\frac{t}{c})}c}{e^{(\frac{-C1}{c})}} + \lambda$$

And it's pretty easy for a human (as opposed to Maple!) to see that this is really a shifted catenary:

$$x(t) = c\,\cosh\left(\frac{t}{c} + d\right) + \lambda,$$

5.11.6. Two Independent Variables. The following procedure handles the case $\iint f(t, s, x(t, s), x_t(t, s), x_s(t, s))\,dt\,ds$:

```
>   EL2indep:=proc(f)
local el1,el2,el3;
el1:=subs({x=x(t,s),dxt=diff(x(t,s),t),dxs=
diff(x(t,s),s)},diff(f,x));
el2:=diff(subs({x=x(t,s),dxt=diff(x(t,s),t),dxs=
diff(x(t,s),s)},diff(f,dxt)),t);
el3:=diff(subs({x=x(t,s),dxt=diff(x(t,s),t),dxs=
diff(x(t,s),s)},diff(f,dxs)),s);
RETURN(simplify(el1-el2-el3)=0);
end:
```

The integral for surface area for a function $z = f(x, y)$ is given by

$$\iint \sqrt{1 + f_x^2 + f_y^2}\,dx\,dy.$$

Keeping with the notation $x(t, s)$, we extremize the integral (searching for a necessary condition for minimum area) to obtain

```
>  EL2indep(sqrt(1+dxt^2+dxs^2));
```

$$-\left(-2\left(\frac{\partial}{\partial t}x(t,\,s)\right)\left(\frac{\partial}{\partial s}x(t,\,s)\right)\left(\frac{\partial^2}{\partial t\,\partial s}x(t,\,s)\right)+\left(\frac{\partial^2}{\partial t^2}x(t,\,s)\right)\right.$$
$$+\left(\frac{\partial^2}{\partial t^2}x(t,\,s)\right)\left(\frac{\partial}{\partial s}x(t,\,s)\right)^2+\left(\frac{\partial^2}{\partial s^2}x(t,\,s)\right)$$
$$+\left.\left(\frac{\partial^2}{\partial s^2}x(t,\,s)\right)\left(\frac{\partial}{\partial t}x(t,\,s)\right)^2\right)\Big/\left(1+\left(\frac{\partial}{\partial t}x(t,\,s)\right)^2\right.$$
$$\left.+\left(\frac{\partial}{\partial s}x(t,\,s)\right)^2\right)^{3/2}$$
$$=0$$

For this equation to be zero, the numerator must be zero. But that then gives the minimal surface equation. Thus, Maple gives a quick derivation of a rather tedious exercise by hand.

5.12. The Brachistochrone

Although the brachistochrone has little to do with minimal surfaces, its important role in the development of the calculus of variations and its amenability to illustration make it an irresistible subject for Maple analysis. The paths of a particle falling under gravity along the cycloid and along a line may be modeled with Maple. Here we shall take the case where the particle always falls from $(0,0)$ and goes to any chosen point. See Example 4.2.3 and Example 4.4.1 for the mathematics involved in proving that the brachistochrone is a cycloid $\alpha(p) = a(p - \sin(p), -1 + \cos(p))$. In order to model the actual motion of the particle along this cycloid, we need to replace the parameter p by time t. The relation between these parameters may be found as follows.

As the particle falls along the cycloid, the parameter p becomes a function of t. Therefore, we can take two derivatives of α with respect to time to get

$$\ddot{\alpha} = a \left[\ddot{p}\sqrt{2}\sqrt{1 - \cos(p)} + \frac{\dot{p}^2}{\sqrt{2}}\sqrt{1 + \cos(p)} \right] T + a \frac{\dot{p}^2}{\sqrt{2}}\sqrt{1 - \cos(p)}\, N,$$

where the tangent and normal vectors respectively are

$$T = \frac{1}{\sqrt{2}}(\sqrt{1 - \cos(p)},\, -\sqrt{1 + \cos(p)}),$$

$$N = \frac{1}{\sqrt{2}}(\sqrt{1 + \cos(p)},\, \sqrt{1 - \cos(p)}).$$

Taking mass $m = 1$, the force of gravity normal to the cycloid is given by $F_{\text{nor}} = g\cos(\theta)$, where θ is the angle between the tangent T and the horizontal. Looking at T's components, we see that $\cos(\theta) = \sqrt{1 - \cos(p)}/\sqrt{2}$. Equating the normal acceleration produced by gravity with the normal component of $\ddot{\alpha}$, we get

$$\frac{g}{\sqrt{2}}\sqrt{1 - \cos(p)} = a\frac{\dot{p}^2}{\sqrt{2}}\sqrt{1 - \cos(p)}$$

$$g = \dot{p}^2 a$$

$$\sqrt{g}\,t = \dot{p}\sqrt{a}, \qquad \text{since } p(0) = 0,$$

$$\sqrt{g/a}\,t = p(t).$$

If we take $g = 1$, then we can write the cycloid in the form

$$x(t) = \frac{at}{\sqrt{a}} - a\sin\left(\frac{t}{\sqrt{a}}\right), \qquad y(t) = -a + a\cos\left(\frac{t}{\sqrt{a}}\right).$$

Exercise 5.12.1. Carry out a similar analysis (but equating tangential components of force with $m = 1$ and acceleration) for a bead sliding down a straight line from $(0,0)$ to a point (A, B) to get the time-parametrization (with $g = 1$)

$$x(t) = \frac{AB}{A^2 + B^2}\frac{t^2}{2}, \qquad y(t) = \frac{B^2}{A^2 + B^2}\frac{t^2}{2}.$$

This now allows us to plot (and animate) the two curves with respect to the same parameter — time:

```
>   with(plots):
```

Now we can plot the cycloid from $(0,0)$ to any chosen point (x,y) (the inputs). Note that the 'if' statements serve to tell 'fsolve' where to look for a solution — before the bottom of the cycloid (i.e. when $p = \pi$) or after.

```
>   cycloidplot:=proc(x,y)
local sol,a,p,x1,p1,cyc;
if (evalf(x) > evalf(Pi)) then x1:=evalf(2*Pi - x)
else x1:=x fi;
sol:=fsolve({a*p-a*sin(p)=x1,-a+a*cos(p)=y},{a,p});
assign(sol);print('the cycloid is',[a*(t-sin(t)),
a*(-1+cos(t))]);
cyc:=[a*(t-sin(t)),a*(-1+cos(t))];
if (evalf(x) > evalf(Pi)) then p1:=evalf(2*Pi - p)
else p1:=p fi;
plot([a*(t-sin(t)),a*(-1+cos(t)),t=0..p1],
ytickmarks=4);
end:
>   cycloidplot(2*Pi-5.7,-1);
```

$$\text{the cycloid is,}$$
$$[.9720784005\, t - .9720784005\sin(t),$$
$$-.9720784005 + .9720784005\cos(t)]$$

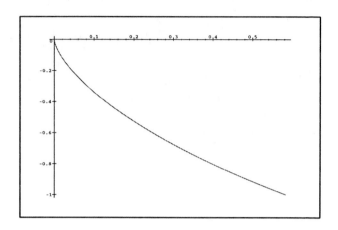

> `cycloidplot(5.7,-1);`

the cycloid is,
$$[.9720784005\, t - .9720784005 \sin(t),$$
$$-.9720784005 + .9720784005 \cos(t)]$$

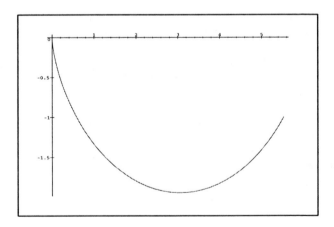

The next procedure plots the line and the cycloid to a chosen point (x, y). The 'S' input allows the user to tell 'fsolve' where to look for a solution for 'p'.

```
>  cyc_line:=proc(x,y,S)
local xx,yy,plot_line,plot_cyc,a,p,end_time_line,sol,
interval;
xx:=-x*y/(x^2+y^2)/2; yy:=-y^2/(x^2+y^2)/2;
end_time_line:=sqrt(-2/y*(x^2+y^2));
plot_line:=plot([xx*t^2,yy*t^2,t=0..end_time_line]):
if (S = 0) then interval:=0..Pi fi;
if (S = 1) then interval:=Pi..2*Pi fi;
sol:=fsolve({a*p-a*sin(p)=x,-a+a*cos(p)=y},{a,p},
p=interval);
assign(sol);
plot_cyc:=plot([a*(t-sin(t)),a*(-1+cos(t)),t=0..p]):
display(plot_cyc,plot_line,scaling=constrained);
end:
```

For the race procedures below, the user must choose 0 or 1 for 'S' telling Maple in which interval to look for a solution to a certain equation. If one of 0 or 1 doesn't work, then the other will. The input 'N' is the number of frames in the animation of the race between the beads sliding down the line and down the cycloid to the chosen point (x, y). Remember, to see an animation, once Maple draws an initial set of axes, click on them and the animation toolbar will appear. The right-pointing triangle starts the animation. So, the 'race' inputs below won't give output here, but they will when the reader runs them.

```
>  race:=proc(x,y,N,S)
local xx,yy,plot_line,plot_cyc,a,p,end_time_line,sol,
interval,s;
xx:=-x*y/(x^2+y^2)/2; yy:=-y^2/(x^2+y^2)/2;
end_time_line:=sqrt(-2/y*(x^2+y^2));
if (S = 0) then interval:=0..Pi fi;
if (S = 1) then interval:=Pi..2*Pi fi;
sol:=fsolve({a*p/sqrt(a)-a*sin(p/sqrt(a))=x,
-a+a*cos(p/sqrt(a))=y},{a,p},p=interval);
assign(sol);
display(seq({pointplot([[xx*(p*s/N)^2,yy*(p*s/N)^2],
[a*p*s/N/sqrt(a)-a*sin(p*s/N/sqrt(a)),-a+
a*cos(p*s/N/sqrt(a))]],symbol=circle)},s=0..N),
insequence=true,xtickmarks=4,ytickmarks=3);
end:
```

```
>  cyc_line(5.7,-1,1);
```

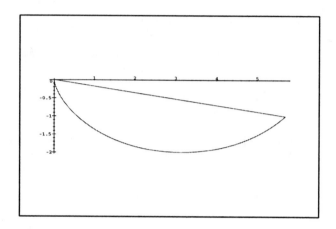

```
>  race(5.7,-1,10,1);
>  cyc_line(2,-1,1);
```

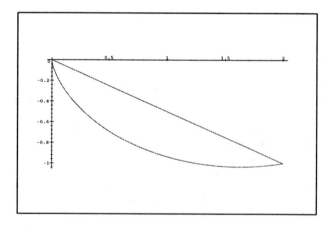

```
>  race(2,-1,10,0);
```

The progress of the beads along the line and cycloid can be compared
by marking the positions at various times and color-coding the positions so that same color denotes same time (modulo a set of colors).
This is what the next procedure does.

```
>  race2:=proc(x,y,N,S)
local xx,yy,plot_line,plot_cyc,a,p,end_time_line,sol,
interval,s,plot_race,col;
col:=array(0..4,[magenta,sienna,black,red,navy]);
xx:=-x*y/(x^2+y^2)/2; yy:=-y^2/(x^2+y^2)/2;
end_time_line:=sqrt(-2/y*(x^2+y^2));
if (S = 0) then interval:=0..Pi fi;
if (S = 1) then interval:=Pi..2*Pi fi;
sol:=fsolve({a*p/sqrt(a)-a*sin(p/sqrt(a))=x,
-a+a*cos(p/sqrt(a))=y},{a,p},p=interval);
assign(sol);
plot_race:=display({seq(pointplot([[xx*(p*s/N)^2,
yy*(p*s/N)^2],[a*p*s/N/sqrt(a)-a*sin(p*s/N/sqrt(a)),
-a+a*cos(p*s/N/sqrt(a))]],symbol=cross,color=
col[s mod 5]),s=0..N)}):
plot_line:=plot([xx*t^2,yy*t^2,t=0..end_time_line],
color=cyan):
plot_cyc:=plot([a*(t/sqrt(a)-sin(t/sqrt(a))),
a*(-1+cos(t/sqrt(a))),t=0..p],color=cyan):
display({plot_race,plot_cyc,plot_line},thickness=2);
end:
>  race2(5.7,-1,10,1);
```

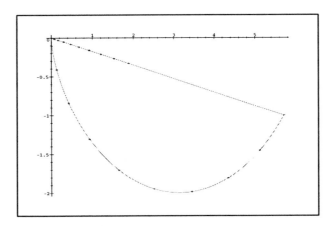

```
>   race2(2,-1,10,0);
```

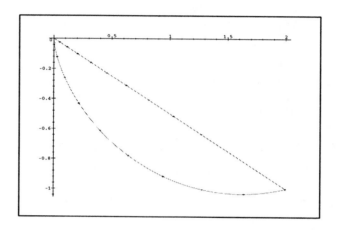

5.13. Surfaces of Delaunay

In this section, we look at the surfaces of revolution which have constant mean curvature. These are the *surfaces of Delaunay*. Recall from Exercise 3.11.2, Exercise 3.11.3 and Exercise 4.5.12 that, for a parametrization $\mathbf{x}(u, v) = (u, h(u) \cos v, h(u) \sin v)$, constant mean curvature is equivalent to

$$h^2 \pm \frac{2ah}{\sqrt{1 + h'^2}} = \pm b^2,$$

and this equation results from solving the problem of minimizing the surface area of a (possibly non-compact) surface of revolution subject to its enclosing a fixed volume. In fact, the equation above is in the form provided by this variational viewpoint, but we shall see that the geometric approach (see Exercise 3.11.3) leads to a better Maple plotting procedure. Nevertheless, we shall first examine surfaces resulting from solutions of the equation above because the analysis of these solutions shows how Maple may be used effectively for such problems.

In this section we use x instead of u and $y(x)$ instead of $h(u)$. Also, we shall restrict ourselves to the case where the $\pm$'s are minus signs. Therefore, we shall be interested in finding the profile curve $y = y(x)$ for the surface of revolution which satisfies the differential equation

$$y^2 - \frac{2ay}{\sqrt{1 + y'^2}} + b^2 = 0.$$

The surface given by the solution of the differential equation above is called an *unduloid*. Of course, the equation can only be solved numerically in general (but see Exercise 3.11.2 for a special case), so this is where Maple can come in. Thanks to John Reinmann for some helpful conversations, especially about premature evaluation (see below).

```
>   with(plots):
```

Let's take particular a and b for concreteness to see where we are heading.

```
>    a:=3;b:=1;
```

$$a := 3$$

$$b := 1$$

When we isolate the derivative, we find that there are two solutions involving square roots (try the 'solve' command below without the 'op' commands). We have to know where the equation is defined, so we need the expression inside the square root of one solution to be non-negative. So, we now isolate the inside of the square root. Each Maple expression is made up of parts called operands, and the 'op' command is used to take the part desired. (There is also a Maple command called 'isolate' which may be used.)

```
>  op(1,op(1,expand(solve(y^2-2*a*y/sqrt(1 +
diff(y(x),x)^2)+b^2=0,diff(y(x),x)))));
```

$$-1 - y^4 + 34\,y^2$$

We can plot this quantity to see where its graph is above the x-axis. Then we can use Maple's 'solve' command to find the actual roots (approximately of course):

```
>  plot(-1-y^4+34*y^2,y=-6..6);
```

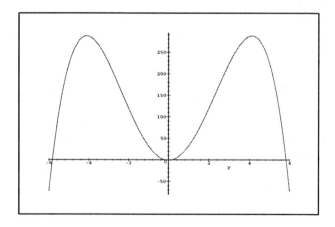

```
>  kkk:=[evalf(solve(-1-y^4+34*y^2=0))];
```

$$kkk := [5.828427124, .171572876, -.171572876, -5.828427124]$$

So now we know that we must stay between about .172 and 5.8. Also, just to see how many operands we have, do the following:

> `nops(kkk);`

4

So now we can write the differential equation with the derivative isolated, and we see that it is, in fact, separable. We then separate variables and plot the values of x as we vary y:

> `eq:=diff(y(x),x)=(-1-y^4+34*y^2)^(1/2)/(y^2+1);`

$$eq := \frac{\partial}{\partial x}\, y(x) = \frac{\sqrt{-1 - y^4 + 34\,y^2}}{y^2 + 1}$$

> `plot(evalf(Int((t^2+1)/sqrt(-1-t^4+34*t^2),`
> `t=0.172..y)),y=0.172..5.8);`

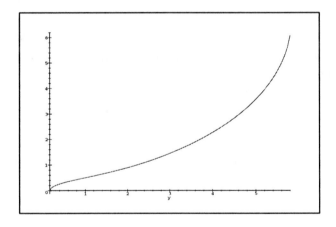

> `evalf(Int(((t^2+1)/sqrt(-1-t^4+34*t^2)),`
> `t=0.172..5.82));`

6.347299174

This number, 6.347299174, is the biggest x can be to get a solution. Otherwise, the corresponding y's give negatives in the square root of the equation. Now we have an idea of the bounds for our ultimate plot. So, let's solve the differential equation numerically with the given initial condition $y(0) = .172$ avoiding a negative square root:

```
>  sol:=dsolve({eq,y(0)=0.172},y(x),numeric,
output=listprocedure):
>  sol(0);
```

$$[\mathrm{x}(0) = 0,\ \mathrm{y}(x)(0) = .172]$$

```
>  sol(6.3);
```

$$[\mathrm{x}(6.3) = 6.3,\ \mathrm{y}(x)(6.3) = 5.817346195911421]$$

Now here is a way to take the numerical solution and plot it. While Maple has commands to plot solutions of differential equations, they aren't useful for us because we eventually want to plot a surface of revolution generated by the numerical curve solution. The next commands substitute the numerical solution into a variable $Y1$ and then create a function Y which has input x and output $Y1(x)$ *if the inputted x is one of the numerical solution x's*. The form of the command prevents Maple from trying to evaluate the function too early. This problem is known to Maple hackers as *premature evaluation*. We can then use Y to graph the surface of revolution desired:

```
>  Y1:=subs(sol,y(x)):
>  Y:=x->if type(x,numeric) then Y1(x) else 'Y'(x) fi:
>  Y(0);
```

$$.172$$

```
>  Y(6.3);
```

$$5.817346195911421$$

To get a picture of the profile curve $y(x)$ (or Y), we take

```
>  plot(Y,0..6.6,scaling=constrained);
```

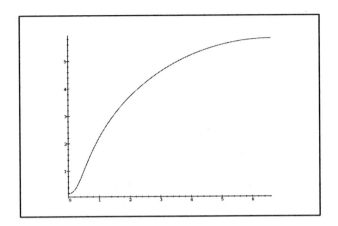

Now, the surface of revolution is non-compact and the profile curves fit together by repetition reminiscent of the Schwarz reflection principle, so we must, by hand, create the reflected images:

```
> X1:=[x,Y(x)*sin(u),-Y(x)*cos(u)];
```

$$X1 := [x,\ Y(x)\sin(u),\ -Y(x)\cos(u)]$$

```
> X2:=[-x,Y(x)*sin(u),-Y(x)*cos(u)];
```

$$X2 := [-x,\ Y(x)\sin(u),\ -Y(x)\cos(u)]$$

```
> X3:=[-12.68+x,Y(x)*sin(u),-Y(x)*cos(u)];
```

$$X3 := [-12.68 + x,\ Y(x)\sin(u),\ -Y(x)\cos(u)]$$

```
> X4:=[12.68-x,Y(x)*sin(u),-Y(x)*cos(u)];
```

$$X4 := [12.68 - x,\ Y(x)\sin(u),\ -Y(x)\cos(u)]$$

Note that we must shift over by $2(6.34) = 12.68$, twice the value of the largest x we can take. We have:

```
> pl1:=plot3d(X1,x=0..6.34,u=0..2*Pi,grid= [10,25]):
> pl2:=plot3d(X2,x=0..6.34,u=0..2*Pi,grid= [10,25]):
> pl3:=plot3d(X3,x=0..6.34,u=0..2*Pi,grid= [10,25]):
> pl4:=plot3d(X4,x=0..6.34,u=0..2*Pi,grid= [10,25]):
```

```
>   display({pl1,pl2,pl3,pl4},scaling= constrained,
style=patch,shading=xy,lightmodel=light3,
orientation=[66,76]);
```

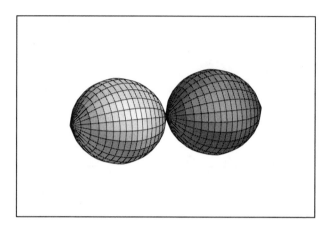

This is the unduloid with two beads. In order to investigate other a's and b's more easily, let's write a procedure to create unduloids.

Exercise 5.13.1. Go through the procedure below and match up elements of the procedure with what was done above.

There is one difference which must be pointed out explicitly in the procedure below. Premature evaluation is handled a different way. Instead of using the 'if' statement as above, we just use '$Y2$'(x) in the definition of the surface parametrization. We do this because MapleV version 4 cannot handle the 'if' statement method inside a procedure. MapleV version 5 *can*, so we give another version of the procedure at the end. The point is that both of these methods of handling premature evaluation are good to keep in mind. Also note that the extra input 'N' below specifies the number of beads ($2N$) to be drawn.

```
>   unduloid:=proc(a,b,N)
local dy,insqrt,sol,k,newsol,i,j,maxnewsol,minnewsol,
maxx,D1,D2,dsol,Y2,W,newy,plotset,M;
dy:=expand(solve(y(x)^2-2*a*y(x)/sqrt(1+diff(y(x),
x)^2)+b^2=0, diff(y(x),x)));print(diff(y(x),x)=dy);
```

```
insqrt:=op(1,op(1,dy));
sol:=[evalf(solve(insqrt=0,y(x)))];
k:=nops(sol);
newsol:=[];
for i from 1 to k do
if (sol[i]>=0) then newsol:=[op(newsol),sol[i]] fi;
od;
maxnewsol:=max(op(newsol))-.0001;
minnewsol:=min(op(newsol))+.0001;
maxx:=evalf(Int(subs(y=t,1/dy),t=minnewsol..
maxnewsol));
print(minnewsol,maxnewsol,maxx);
dsol:=dsolve({diff(y(x),x)=dy,y(0)=minnewsol},y(x),
numeric, output=listprocedure);
Y2:=subs(dsol,y(x));
plotset:={};M:=1-N;
for j from M to N do
D1:=[x-2*j*maxx,'Y2'(x)*sin(v),-'Y2'(x)*cos(v)];
D2:=[-x+2*j*maxx,'Y2'(x)*sin(v),-'Y2'(x)*cos(v)];
plotset:={op(plotset),D1,D2};
od;
plot3d(plotset,x=0..maxx,v=0..2*Pi,scaling=
constrained,grid=[10,25],style=patch,shading=xyz,
lightmodel=light3,orientation=[66,76]);
end:
> unduloid(1.5,1,1);
```

$$\frac{\partial}{\partial x}\,y(x) = \frac{\sqrt{-1 - y(x)^4 + 7\,y(x)^2}}{y(x)^2 + 1.}$$

.382066011,   2.617933989,   3.929731659

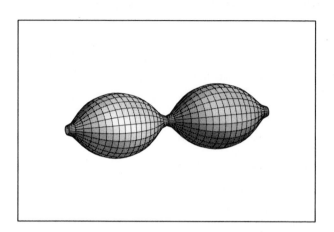

```
> unduloid(1.5,1,2);
```

$$\frac{\partial}{\partial x}\,\mathrm{y}(x) = \frac{\sqrt{-1 - \mathrm{y}(x)^4 + 7\,\mathrm{y}(x)^2}}{\mathrm{y}(x)^2 + 1.}$$

$$.382066011, \quad 2.617933989, \quad 3.929731659$$

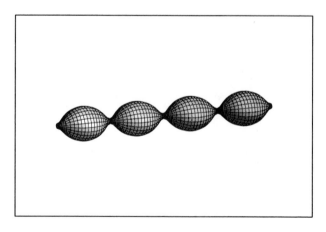

```
> unduloid(3,1,1);
```

$$\frac{\partial}{\partial x}\,\mathrm{y}(x) = \frac{\sqrt{-1 - \mathrm{y}(x)^4 + 34\,\mathrm{y}(x)^2}}{\mathrm{y}(x)^2 + 1}$$

.171672876,   5.828327124,   6.641252049

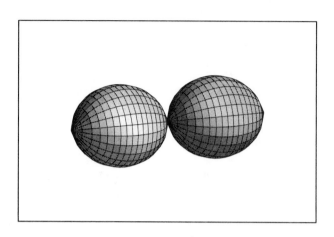

```
>  unduloid(1.5,1.499,1);
```

$$\frac{\partial}{\partial x}\, \mathrm{y}(x) =$$

$$\frac{\sqrt{-5049013494001 - 1000000000000\, \mathrm{y}(x)^4 + 4505998000000\, \mathrm{y}(x)^2}}{.1000000\,10^7\, \mathrm{y}(x)^2 + .2247001\,10^7}$$

1.445336874,   1.554663126,   4.529523125

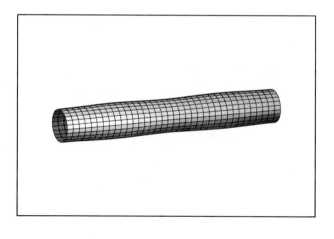

So, as a and b get close, the unduloid becomes a cylinder in accordance with its description as a roulette of a conic (see any of [Del41], [Eel87] or [Opr97] for instance). Finally, here is the MapleV version 5 procedure:

```
>   unduloid5:=proc(a,b,N)
local dy,insqrt,sol,k,newsol,i,j,maxnewsol,minnewsol,
maxx,D1,D2, dsol,Y2,W,newy,plotset,M;
dy:=expand(solve(y(x)^2-2*a*y(x)/sqrt(1+diff(y(x),
x)^2)+b^2=0, diff(y(x),x))));print(diff(y(x),x)=dy);
insqrt:=op(1,op(1,dy));
sol:=[evalf(solve(insqrt=0,y(x)))];
k:=nops(sol);
newsol:=[];
for i from 1 to k do
if (sol[i]>=0) then newsol:=[op(newsol),sol[i]] fi;
od;
maxnewsol:=max(op(newsol))-.0001;
minnewsol:=min(op(newsol))+.0001;
maxx:=evalf(Int(subs(y=t,1/dy),t=minnewsol..
maxnewsol));
print(minnewsol,maxnewsol,maxx);
dsol:=dsolve({diff(y(x),x)=dy,y(0)=minnewsol},y(x),
numeric, output=listprocedure);
Y2:=subs(dsol,y(x));
W:= x->if type(x,numeric) then Y2(x) else 'W'(x) fi;
plotset:={ };M:=1-N;
for j from M to N do
D1:=[x-2*j*maxx,W(x)*sin(v),-W(x)*cos(v)];
D2:=[-x+2*j*maxx,W(x)*sin(v),-W(x)*cos(v)];
plotset:={op(plotset),D1,D2};
od;
plot3d(plotset,x=0..maxx,v=0..2*Pi,scaling=
constrained,grid=[10,25], style=patch,shading=xyz,
lightmodel=light3,orientation=[66,76]);
end:
```

Exercise 5.13.2. Use the formulation of the Delaunay equation given in Exercise 3.11.3 to justify the procedure below which plots *all* the surfaces of Delaunay. This procedure is essentially due to J. Reinmann, as is the overall approach. The inputs to the procedure are H (mean curvature), c (a constant), N (the number of nodes of the surface), *angle* (the total revolution angle), *gr* (the grid size) and *orient* (the orientation of the plot). The input *angle* is needed for the nodoid because a full plot with $angle = 2\pi$ covers up the inside self-intersecting pieces.

```
>   delaunsurf:=proc(H,c,N,angle,gr,orient)
local du,start,uend,i,j,hlims,plot1,plot2,dsol,u2,
plotset,ending;
hlims:=sort([fsolve(h^2-(H*h^2-c)^2=0,h)]);
start:=hlims[1]+0.00001;
ending:=hlims[2]-0.00001;
uend:=evalf(Int(((H*t^2-c)/sqrt(t^2-(H*t^2-c)^2)),
t=start..ending));
du:=(H*h^2-c)/sqrt(h^2-(H*h^2-c)^2);
dsol:=dsolve({diff(u(h),h)=du,u(start)=0},u(h),
numeric,output=listprocedure);
u2:=subs(dsol,u(h));
for j from 1 to N do
plot1[j]:=plot3d([h*cos(v),(j-1)*2*uend+'u2'(h),
h*sin(v)],h=start..ending,v=0..angle,grid=gr);
plot2[j]:=plot3d([h*cos(v),j*2*uend-'u2'(h),
h*sin(v)],h=start..ending,v=0..angle,grid=gr);
od;
display(seq({plot1[k],plot2[k]},k=1..N),scaling=
constrained,style=patch,shading=xy,lightmodel=
light2,orientation=orient);
end:
```

Here is a case where H and c have the same sign and we obtain the nodoid:

```
> delaunsurf(1,0.5,2,2*Pi/3,[15,20],[45,45]);
```

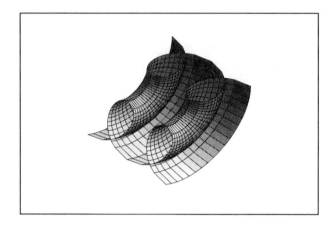

```
> delaunsurf(1,0.5,2,12*Pi/7,[20,16],[-139,76]) ;
```

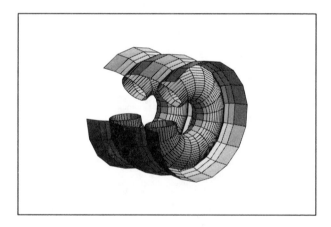

To see the full plot of the nodoid, input the command below:

```
> delaunsurf(1,0.5,2,2*Pi,[20,25],[-139,76]);
```

Here is the case where $c = 0$. This confirms what you should have obtained in Exercise 3.11.3.

```
>  delaunsurf(2,0,2,2*Pi,[20,25],[-180,76]);
```

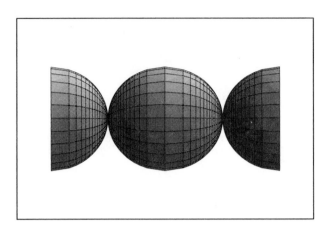

Now let's see what happens when H and c have opposite signs, but obey $-1/4 < Hc$:

```
>  delaunsurf(0.7,-0.22,2,2*Pi,[15,20],[-139,76] );
```

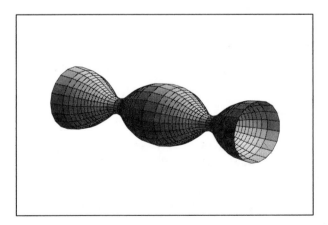

As we get close to the magic $-1/4$, the unduloids become a cylinder:

```
>  delaunsurf(1,-0.24999,2,2*Pi,[6,20],[-139,76] );
```

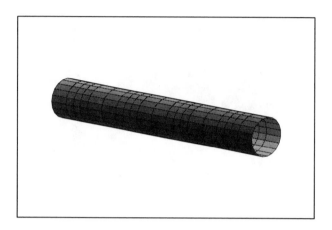

To end this section, we give a procedure for a constant mean curvature surface which is not a surface of revolution. This surface is called Sievert's surface. The definition of the parametrization is complicated enough to warrant a procedure, so the reader is warned not to verify constant mean curvature by using the MK procedure of § 5.5.

```
>  with(plots):with(linalg):
```

In the following, the input B is a real non-zero number:

```
>  sievert:=proc(B)
local a,b,denom,m,X;
a:=sinh(B)*u;
b:=cosh(B)*v;
denom:=sinh(B)*((cosh(2*a)-cos(2*b))*cosh(2*B)+
(2+cosh(2*a)+cos(2*b)));
m:=evalm(cosh(B)*[sinh(a),sin(b)*cos(v),
sin(b)*sin(v)] + [0,-cos(b)*sin(v),cos(b)*cos(v)]);
X:=evalm([u,0,0]-(8*cosh(B)*cosh(a)/denom)*m);
end:
```

```
>  plot3d(sievert(0.75),u=-2.5..2.5,v=-10.5..10.5,
scaling=constrained,grid=[30,100],style=patch,
shading=xy,lightmodel=light3,orientation=[-3,140]);
```

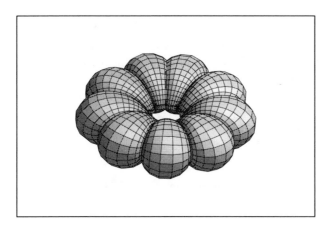

5.14. The Mylar Balloon

From Example 4.5.13, we saw that the shape of a mylar balloon may
be obtained via variational analysis.   More specifically, the shape of
a mylar balloon is the solution to the problem of maximizing volume
subject to having a fixed profile curve length. Here we will try to
visualize the balloon and calculate its inflated radius, thickness and
volume in terms of the radius of the original disks used to make the
balloon.

```
> with(plots):
```

The following is a procedure which takes as input the inflated radius
of the balloon and which gives as output a picture of the balloon.
Notice that we again use the trick of avoiding premature evaluation
as in § 5.13:

```
> mylar:=proc(a)
local X,X1,desys,hu,p1,p2;
X:=[u*cos(v),u*sin(v),h];
X1:=[u*cos(v),u*sin(v),-h];
desys:=dsolve({diff(h(u),u)=-u^2/sqrt(a^4-u^ 4),
h(a-0.0001)=0.00001}, h(u),type=numeric,output=
listprocedure):
hu:=subs(desys,h(u));
p1:=plot3d(subs(h='hu'(u),X),u=0..a-0.0001,v= 0..2*Pi):
p2:=plot3d(subs(h='hu'(u),X1),u=0..a-.0001,v=0..2*Pi):
display({p1,p2},scaling=constrained,style=patch,
shading=zhue,orientation=[84,82],axes=boxed);
end:
> mylar(2);
```

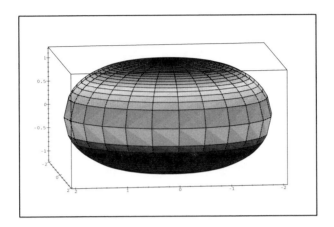

Recall that the disk radius r is related to the inflated radius a by

$$r = \frac{a}{4} \int_0^1 \frac{t^{-3/4}}{\sqrt{1-t}}\, dt,$$

while the thickness of the balloon is given by

$$\text{thickness} = \frac{a}{2} \int_0^1 \frac{t^{-1/4}}{\sqrt{1-t}}\, dt$$

and the volume by

$$V = \frac{\pi a^3}{2} \int_0^1 \frac{u^{1/4}}{\sqrt{1-u}}\, du.$$

Let's calculate these quantities here:

```
>   r=a/4*evalf(Int(t^(-3/4)/sqrt(1-t),t=0..1));
```

$$r = 1.311028777\, a$$

```
>   a=1/1.311028777*r;
```

$$a = .7627597636\, r$$

The inflated radius is then about 0.763 of the original. The thickness is then given by

```
> twice_y(0)=a/2*evalf(Int(t^(-1/4)/sqrt(1-t),
t=0..1));
thickness=.7627597636*r/2*evalf(Int(t^(-1/4)/
sqrt(1-t), t=0..1));
```

$$twice\_y(0) = 1.198140235\,a$$

$$thickness = .9138931620\,r$$

Finally, the volume is calculated to be

```
> volume=a^3*evalf(Int(Pi/2*u^(1/4)*1/sqrt(1-u),
u=0..1));
volume=(.7627597636*r)^3*evalf(Int(Pi/2*u^(1/4)*1/
sqrt(1-u),u=0..1));
```

$$volume = 2.745812250\,a^3$$

$$volume = 1.218524217\,r^3$$

as compared to the volume of the sphere of radius a,

```
> aspherevol=evalf(4/3*Pi)*(.7627597636*r)^3;
```

$$aspherevol = 1.858882486\,r^3$$

Bibliography

[AT76]   F. Almgren and J. Taylor, *The geometry of soap films and soap bubbles*, Scientific American **235** (July 1976), 82–93.

[BC86]   J. Barbosa and A. Colares, *Minimal Surfaces in R^3*, Lecture Notes in Math., vol. 1195, Springer-Verlag, 1986.

[Bli46]   G. Bliss, *Lectures on the Calculus of Variations*, U. of Chicago Press, 1946.

[Boy59]   C. V. Boys, *Soap Bubbles: Their Colors and the Forces which Mold Them*, Dover, 1959.

[Can70]   P. B. Canham, *The minimum energy of bending as a possible explanation of the biconcave shape of the human red blood cell*, J. Theoret. Biol. **26** (1970), 61–81.

[CH53]   R. Courant and D. Hilbert, *Methods of Mathematical Physics*, vol. 1, Interscience, 1953.

[Che55]   S. S. Chern, *An elementary proof of the existence of isothermal parameters on a surface*, Proc. Amer. Math. Soc. **6** (1955), 771–782.

[Cou50]   R. Courant, *Dirichlet's Principle, Conformal Mapping and Minimal Surfaces*, Interscience, 1950.

[CR43]   R. Courant and H. Robbins, *What is Mathematics*, Oxford U. Press, 1943.

[dC76]   M. do Carmo, *Differential Geometry of Curves and Surfaces*, Prentice-Hall, 1976.

[Del41]   C. Delaunay, *Sur la surface de revolution dont la courbure moyenne est constante*, J. de Math. Pures et Appl. **6** (1841), 309–320.

[DH76a]   H. Deuling and W. Helfrich, *The curvature elasticity of fluid membranes: a catalogue of vesicle shapes*, Le Journal de Physique **37** (1976), 1335–1345.

[DH76b]   H. Deuling and W. Helfrich, *Red blood cell shapes as explained on the basis of curvature elasticity*, Biophys. J. **16** (1976), 861–868.

[DHKW92]   U. Dierkes, S. Hildebrandt, A. Küster, and O. Wahlrab, *Minimal Surfaces I*, Grundlehren der Math. Wiss., vol. 295, Springer-Verlag, 1992.

[Dou31]   J. Douglas, *Solution of the problem of Plateau*, Trans. Amer. Math. Soc. **33** (1931), 263–321.

[Eel78]   J. Eells, *On the surfaces of Delaunay and their Gauss maps*, Proc. IV Int. Colloq. Diff. Geom., Cursos Congr. Univ. Santiago de Compostela **15** (1978), 97–116.

[Eel87]   J. Eells, *The surfaces of Delaunay*, Math. Intell. **9 no. 1** (1987), 53–57.

[Ewi85]   G. Ewing, *Calculus of Variations with Applications*, Dover, 1985.

[Fin86]   R. Finn, *Equillibrium Capillary Surfaces*, Springer-Verlag, 1986.

[For68]   M. Forray, *Variational Calculus in Science and Engineering*, McGraw-Hill, 1968.

[FT91]   A. Fomenko and A. Tuzhilin, *Elements of the Geometry and Topology of Minimal Surfaces in Three-Dimensional Space*, Transl. of Math. Mono., vol. 93, Amer. Math. Soc., 1991.

[Gar95]   R. Gardner, *Experiments with Bubbles*, Enslow Publishers Inc., 1995.

[GF63]   I. M. Gelfand and S. V. Fomin, *Calculus of Variations*, Prentice-Hall, 1963.

[Gra40]   W. Graustein, *Harmonic minimal surfaces*, Trans. Amer. Math. Soc. **47** (1940), 173–206.

[Gra93]   A. Gray, *Modern Differential Geometry of Curves and Surfaces*, CRC Press, 1993.

[HHS95]   J. Hass, M. Hutchings, and R. Schlafly, *The double bubble conjecture*, Elect. Res. Announce. AMS **1 no. 3** (1995), 98–102.

[HK95]   L. Haws and T. Kiser, *Exploring the brachistochrone problem*, Amer. Math. Monthly **102 no.4** (1995), 328–336.

[HM90]   D. Hoffman and W. Meeks, *Minimal surfaces based on the catenoid*, Amer. Math. Monthly **97 no. 8** (1990), 702–730.

[Hof87]   D. Hoffman, *The computer-aided discovery of new embedded minimal surfaces*, Math. Intell. **9 no. 3** (1987), 8–21.

[HT85]   S. Hildebrandt and A. Tromba, *Mathematics and Optimal Form*, Scientific American Books, 1985, rev. ed., *The Parsimonious Universe*, Copernicus, 1996.

[Ise92]   C. Isenberg, *The Science of Soap Films and Soap Bubbles*, Dover, 1992.

[Lap05]   P. S. Laplace, *Mechanique Celeste*, vol. IV, Duprat, 1805, English transl., Hilliard, Gray, Little and Wilkins, 1839; reprint, Chelsea Publ. Co., 1966.

[Law96]   G. Lawlor, *A new minimization proof for the brachistochrone*, Amer. Math. Monthly **103 no. 3** (1996), 242–249.

[Lem97]   D. Lemons, *Perfect Form*, Princeton University Press, 1997.

[LF51]   L. Lusternik and A. Fet, *Variational problems on closed manifolds*, Dokl. Akad. Nauk. SSSR **81** (1951), 17–18, (Russian).

[Lov94]   D. Lovett, *Demonstrating Science with Soap Films*, Institute of Physics Publ., 1994.

[Max49]   J. Maxwell, *On the theory of rolling curves*, Trans. Roy. Soc. Edinburgh **XVI Part V** (1849), 519–544.

[MH87]   J. Marsden and M. Hoffman, *Basic Complex Analysis*, W. H. Freeman, 1987.

[Mor88]   F. Morgan, *Geometric Measure Theory: A Beginner's Guide*, Academic Press, 1988.

[Mor92]   F. Morgan, *Minimal surfaces, crystals, shortest networks, and undergraduate research*, Math. Intell. **14 no. 3** (1992), 37–44.

[MT88]   J. Marsden and A. Tromba, *Vector Calculus*, W. H. Freeman, 1988.

[Nit89]   J. Nitsche, *Lectures on Minimal Surfaces*, vol. 1, Cambridge U. Press, 1989.

[Opr97]   J. Oprea, *Differential Geometry and its Applications*, Prentice-Hall, 1997.

[Oss86]   R. Osserman, *A Survey of Minimal Surfaces*, Dover, 1986.

[Oss90]   R. Osserman, *Curvature in the Eighties*, Amer. Math. Monthly **97 no. 8** (1990), 731–756.

[Pau94]   W. Paulsen, *What is the shape of a mylar balloon*, Amer. Math. Monthly **101 no. 10** (1994), 953–958.

[PS93]   U. Pinkall and I. Sterling, *Computational aspects of soap bubble deformations*, Proc. Amer. Math. Soc. **118 no. 2** (1993), 571–576.

[Rad71]   T. Radó, *On the Problem of Plateau/Subharmonic Functions*, Springer-Verlag, 1971.

[Ros88]   A. Ros, *Compact surfaces with constant scalar curvature and a cogruence theorem*, J. Diff. Geom. **27** (1988), 215–220.

[Sag92]   H. Sagan, *Introduction to the Calculus of Variations*, Dover, 1992.

[Spi79]   M. Spivak, *A Comprehensive Introduction to Differential Geometry: 5 volumes*, Publish or Perish, 1979.

[SS93]   E. Saff and A. Snider, *Fundamentals of Complex Analysis for Mathematics, Science and Engineering*, Prentice-Hall, 1993.

[Str88]   D. Struik, *Lectures on Classical Differential Geometry*, Dover, 1988.

[SZ47]   F. Sears and M. Zemansky, *College Physics*, Addison-Wesley, 1947.

[Tay76]   J. Taylor, *The structure of singularities in soap-bubble-like and soap-film-like minimal surfaces*, Ann. Math. **103** (1976), 489–539.

[TF91]   D. T. Thi and A. T. Fomenko, *Minimal Surfaces, Stratified Multivarifolds and the Plateau Problem*, Transl. of Math. Monographs, vol. 84, Amer. Math. Soc., 1991.

[Tho92]   D. W. Thompson, *On Growth and Form: The Complete Revised Edition*, Dover, 1992.

[Tik90]   V. M. Tikhomirov, *Stories about Maxima and Minima*, Math. World, vol. 1, Amer. Math. Soc., 1990.

[Tro96]   J. Troutman, *Variational Calculus and Optimal Control*, Undergrad. Texts in Math., Springer-Verlag, 1996.

[Wei74]   R. Weinstock, *Calculus of Variations*, Dover, 1974.

[Wen86]   H. Wente, *Counter-example to the Hopf conjecture*, Pac. J. Math. **121** (1986), 193–244.

[You05]   T. Young, *An essay on the cohesion of fluids*, Phil. Trans. Roy. Soc. (London) **1** (1805), 65–87.

Index